D1805840

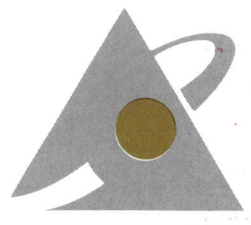

MATERIALS

BATH SCIENCE 5-16

£5.50

Nelson

Thomas Nelson and Sons Ltd
Nelson House Mayfield Road
Walton-on-Thames Surrey
KT12 5PL UK

51 York Place
Edinburgh
EH1 3JD UK

Thomas Nelson (Hong Kong) Ltd
Toppan Building 10/F
22A Westlands Road
Quarry Bay Hong Kong

Thomas Nelson Australia
102 Dodds Street
South Melbourne
Victoria 3205 Australia

Nelson Canada
1120 Birchmount Road
Scarborough Ontario
M1K 5G4 Canada

© Thomas Nelson and Sons Ltd

First published by Thomas Nelson and Sons Ltd

ISBN 0-17-438433-5
NPN 9 8 7 6 5 4 3 2 1

Designed by Fox and Partners, Bath

Printed in Great Britain

The Project Team

The *Bath Science* Project Team is directed by Professor Jeff Thompson CBE and Dr Martin Hollins of the School of Education, University of Bath. The authors for the Key Stage 4 materials were:

John Collins   John Harris   Christine Harrison   Martin Hollins
Theresa Kinnison   Janet Major   David Sang

with editorial supervision by Janet Major, Christine Harrison and Martin Hollins. *Materials* was written by Janet Major.

Acknowledgements

The authors and publishers wish to acknowledge with thanks, the following photographic sources: Aerofilms Ltd p. 74; Barnaby's Picture Library pp. 6 right, 53, 109 centre; BASF, Germany p. 55 bottom; Anthony Blake Photo Library p. 17; Bodleian Library, Oxford p. 69; British Coal p. 25; British Petroleum p. 48; Büchi Laboratorium p. 6 top; J. Allan Cash Photo Library pp. 49 top left, 67; Del Monté p. 19; Mary Evans Picture Library p. 103 right; Robert Harding Picture Library pp. 37, 41, 62, 92 left, 105 top; Michael Holford p. 103 top; Hulton Picture Library pp. 8 bottom, 11 left, 30 left, 44, 55 top, 85; Imperial Chemical Industries pp. 30 right, 56; La Farge Special Cements p. 22; Mansell Collection p. 26 right; Metropolitan Police p. 80 left; Museum of Science and Industry p. 39 bottom; Picturepoint Ltd pp. 57, 76, 104 bottom, 106 top, 109 left, 109 top right; Popperfoto p. 43; Valerie Randall pp. 50, 59 top, 64, 68, 75, 92 right, 102, 109 bottom right; Ann Ronan Picture Library pp. 60, 81 bottom; Royal Society of Great Britain p. 34; Sarsons/The Nestlé Company p. 15 right; Science Photo Library pp. 4, 7 left, 8 top, 10, 27, 33, 36, 37, 59 centre, 80 right, 84, 86, 90, 97, 101, 103 bottom, 108; Alan Thomas pp. 7 right, 9 top, 11 right, 13, 14, 18, 31, 50, 81 top, 94, 95; Tony Stone (Worldwide) *Chemical Pathways* contents page, pp. 3, 42; University of Cambridge, Cavendish Laboratory pp. 70 right, 71; Zefa Picture Library pp. 9 bottom, 49 right.

Illustrations drawn by: Sarah Mabbutt pp. 4 centre & right, 6, 18 left, 26, 29 left, 38, 47 left, 50, 51 right, 52 right, 54, 59, 63 right, 67, 71, 72, 74, 75, 76, 77 top, 78, 79 bottom, 87 bottom, 90, 91 right, 92, 93, 94, 100, 107; Dave Bowyer pp. 4 bottom, 5 left, 12 top, 13 left, 14, 16, 19, 20, 22, 35, 49; Miranda Gray pp. 5 right, 15, 18 right, 24 bottom, 41, 53; Linden Artists/Craig Warwick pp. 9, 28, 29 right, 51 left, 52 left, 61 bottom, 79 top, 85, 87 top, 95; Edna A. Moore pp. 10, 13 top, 17, 21 right, 27, 30, 31, 32, 36, 37, 40, 45, 46, 47 right & bottom, 61 top, 62, 63 left, 66 centre, 68 bottom, 77 bottom, 83, 84, 88, 89, 91 left, 99, 106, 108; Jolyon Webb pp. 12 bottom, 13 right, 21 left, 86; Steve Noon pp. 24 top, 44, 58, 66 top, 68 centre, 98, 110, 111; Fox and Partners pp. 25, 64, 102; Lynn Williams p. 33 left; David Pow p. 82; Brian Walker p. 104.

The publishers have made every effort to trace the copyright holders, but where they have failed to do so they will be pleased to make the necessary arrangements at the first opportunity.

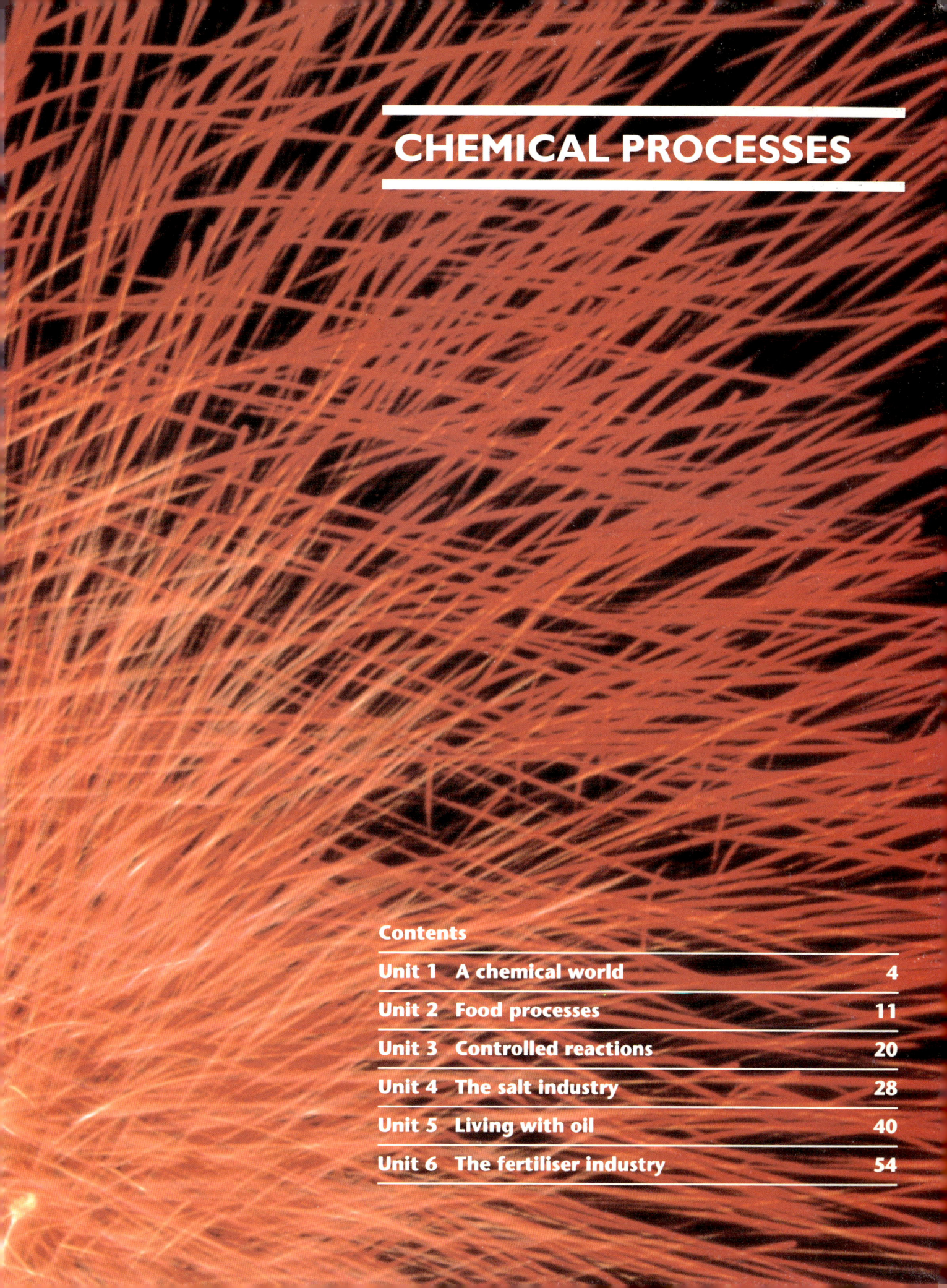

CHEMICAL PROCESSES

Contents

UNIT 1
A CHEMICAL WORLD

This impressive photograph of our planet Earth was taken from Apollo-8 on Christmas Day, 1968

Contents

oxygen	65%
carbon	18%
hydrogen	10%
nitrogen	3%
calcium	2%
phosphorus	1%
others	1%

Contents

oxygen	89%
hydrogen	11%

What could these labels be from? You have probably seen these objects today!

A. Discuss

- Discuss the meaning of the word chemical and agree on a definition that begins like this: 'A chemical substance is....

- These three statements were made by people in a waiting room at a doctor's surgery:

  It's all those chemicals they use. They're all bad.

  I won't let my children have any E numbers.

  I'm just going to the chemists to get my prescription.

What do the statements imply?

Section 1.1

WHAT IS A CHEMICAL?

Our planet and its surrounding atmosphere are made up entirely of chemicals. Everything that we see, touch, taste or smell consists of chemicals. You have probably seen, heard and used the word 'chemical' lots of times. But, have you ever thought what it means?

Wherever you are, you are surrounded by chemical substances. The clothes you wear, the food you eat and the houses you live in are all made up of chemical substances. Many different materials are used in building, decorating and furnishing a house. Each of these materials is made from one or more chemical substances.

1.30–3.30	ORGANIC CERAMICS FOR CHRISTMAS	Course fee £32.50
	Bring out the natural beauty of clay without using chemicals	
7.00–9.00	FRENCH FOR BEGINNERS	Course fee £25.00

What is meant by the word 'chemical' in each of these illustrations?

B. Research

- Choose a building which you know. It could be the house where you live.

- Do a materials survey of the inside and outside of the building.

- Identify as many as possible of the chemical substances that make up these materials.

- Present your information in the most interesting way (table, diagram, leaflet, poster etc.).

WHAT DO WE MEAN BY PURE?

The picture shows three different types of salt. They have different names. They are intended for different uses.

salt
anti-caking agent
535

salt
anti-caking agents
sodium hexacyanoferrate(II)
magnesium carbonate

sea salt
sodium iodide
magnesium carbonate

C. Work out

- Make a list of the ways in which the three types of salt shown in the picture are different from each other.

  Which of them, if any, is pure salt? Explain your answer.

D. Discuss

- Pure substances are one type of chemical substance. Look carefully at the collection of substances you have been given. Decide which ones, if any, are pure.

It is not easy to sort out pure substances from other substances just by looking at them. For example, you might think that antiseptic cream is pure – after all you use it on your skin. In fact, it is a mixture made up of lots of different substances. One well known antiseptic cream contains the following ingredients: anhydrous lanolin, yellow soft paraffin, light liquid paraffin, starch, zinc oxide, methylsalicylate, octaphonium chloride, chloroxylenol, phenol and menthol!

Aspirin has been called a a wonder drug. It has antipyretic (fever reducing), analgesic (pain killing) and anti-inflammatory (swelling reducing) properties. Research has also shown that aspirin may help to prevent blood clots forming and be effective in treating diseases like heart attacks and strokes.

The aspirin story began in 1763 when Edward Stone, a clergyman, was trying to find a cure for malaria. He suggested that the bark of the willow tree might provide this cure because willow trees grew in the damp areas where malaria was common. In those days, many people believed that a cure for a disease could be found where the disease was prevalent. Stone found that the extract from the bark did have antipyretic properties but it did not cure malaria. Doctors wanted to know exactly what this antipyretic substance was. The extract from the bark was a mixture of chemical substances. Scientists worked hard to separate out the substance that reduced the fever. Only when the pure substance was isolated could scientists be certain which part of the bark was the 'active ingredient'. Over 100 years later this chemical substance was identified as acetylsalicyclic acid (or 2-ethanoylhydroxy-benzoic acid). Once it had been identified it became possible to make it synthetically. The first synthetic aspirin was made in 1899.

How do we know when a substance is pure? A pure substance has a sharp melting point. It also has a sharp boiling point. Scientists can get evidence on how pure substances are by looking at their melting and boiling points.

E. Test and report

- Imagine that you are working in the quality control department of a factory making aspirin. Two samples have been brought to you to test. From previous work you know that the melting point of pure aspirin is 136°C. Carry out a melting point test on each of the samples. Worksheet CPr1 gives details of how to do this.

- Write a brief report explaining your procedure and conclusions. Be prepared to present your report to the rest of the class.

SALIX ALBA
White Willow   Salicaceae
(Europe Asia)

Since the time of Hippocrates, the ancient Greek 'Father of Medicine', willow extracts have been used for remedies of illnesses.

MAKING THINGS PURER

This magnificent ruby is on the front of the Imperial State crown

Electrical apparatus for determining the melting points of substances

It would be too expensive in terms of labour to have a person doing the kind of melting point test you have just done on each batch of chemical substances made at a factory. The photograph shows the kind of equipment that is used in some factories for testing the purity of a product. The product is heated electrically.

The ruby shown in this photograph was formed inside the Earth many millions of years ago. At that time the Earth's crust was very hot and there was a lot of volcanic activity. Much of the rock of the Earth's crust was molten. Molten rock is a solution of many different substances. As the Earth cooled, the substances in the molten rock turned into crystals. The ruby was formed during this process of crystallisation. The process of crystallisation can be used to purify chemical substances. In the next activity you can use this method to purify an impure sample of aspirin.

Batch 6	
Sample 1	136.00
Sample 2	136.00
Sample 3	135.99
Sample 4	135.05
Sample 5	135.04
Sample 6	134.00
Sample 7	134.00
Sample 8	133.00
Sample 9	132.00
Sample 10	131.00

F. Interpret   EXTENSION

- Look at the print out showing the results of the melting point tests carried out on ten different samples of aspirin. If you were in control of these tests:

1 What action would you take immediately and why?

2 Which sample (or samples) would you send to a customer?

3 What would you do with the impure samples?

G. Observe and record

- Put about two spatulas of the contaminated aspirin into a boiling tube.

- Add ethanoic acid to quarter fill the tube.

- Heat the tube gently until the aspirin has dissolved.

- Pour the hot solution through a filter into a clean boiling tube.

- Allow the filtrate to cool. The aspirin will crystallise.

- Pour the filtrate through a clean filter. The crystallised aspirin will remain on the filter paper.

- Carry out a melting point test on the dried aspirin to check how pure it is.

- Draw a series of labelled diagrams to show how you purified the aspirin.

Producing a new drug takes many years and is very expensive. A lot of time and money is spent on research and development. These are some of the questions that need to be considered:

What is the target disease?	How effective is the drug in treating the disease?	Can the company finance the necessary development of the drug?
What are the long-term side effects on the body?	How does the body absorb the drug?	How long does the drug last in the body?
Is any treatment already available for the target disease?	Does production of the drug fit into the future planning of the company?	What are the breakdown products of the drug in the body, and what happens to them?
Will the company make a profit?	Where does the drug build up in the body?	
What are the short-term side effects on the body?		

Some of the stages in the production of a new drug

Research

⬇

Development

- Initial clinical trials on volunteers
- Tests on human patients
- Full clinical trials on patients
- Long term clinical trials

The development of safe and effective drugs is a combined operation between chemists and doctors. The manufacture of drugs is subject to stringent controls. Any material which fails to reach a specified level of purity has to be discarded. As a further check on safety, doctors continuously monitor the use of drugs, assessing their effectiveness and watching for side effects.

H. Work out

- Look through the list of questions that need to be answered during the research and development stages of a new drug. Decide which ones must be answered in the research phase and which ones can only be answered in the clinical trials.

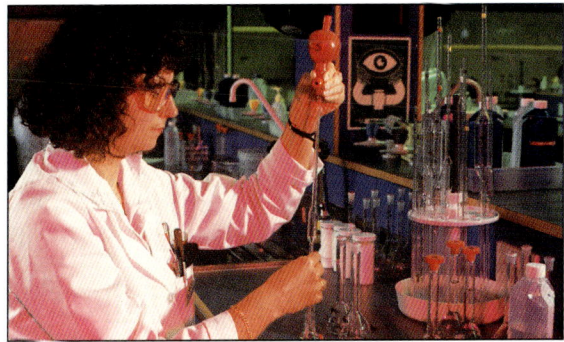

Quality control tests being carried out in a pharmaceutical laboratory

ELEMENTS OR COMPOUNDS

A chemical compound consists of two or more simple substances called elements which are chemically combined.

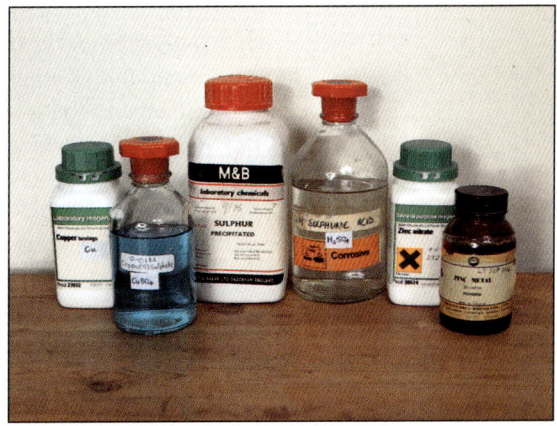

How can we tell if a substance is an element?

I. Discuss

- Look at each of the substances in the photograph and decide whether it is an element or a compound.

  How did you do this?

  What other information can you get from the labels?

It is possible to tell if a substance is an element or a compound if we know what its chemical formula is. For example, $CuSO_4$ is the chemical formula of copper(II) sulphate. This formula tells us, in chemical shorthand, that copper(II) sulphate is a compound containing the elements copper, sulphur and oxygen. The substance with the chemical formula Cu just contains the element copper.

Recognising that all substances are made up from elements is not a new idea. The early Greeks thought that there were some basic elements (simple substances) from which everything else was made. Their theory was that only four elements existed: earth, air, fire and water.

Much later, in the Middle Ages, the alchemists of Europe and the Middle East refined this theory and came up with a new system of three first principles. These were named salt, sulphur and mercury. It was Robert Boyle, an Irish experimental philosopher who wrote the Sceptical Chemist in 1677, who made a statement which changed the way people thought about elements. He said:

'If A is united to B the result C is definitely a compound. But whether A is to be called an element depends on whether it is possible to take anything out of it which is different from the original A, leaving a residue which is also different from A. Also, if we take a substance A and we manage to get new substances from it, for example, A → B+C, we know that A is not an element. But, if A does not form any new substances in our experiments then we can not say definitely that A is not an element. Perhaps it is a compound that is difficult to split up.'

This idea was so useful to scientists that Boyle became known as the Father of Chemistry.

Collecting samples from a crater on the moon

Robert Boyle 1627 – 1691

Unfortunately, when a new substance is discovered it does not come with its chemical formula attached to it on a label. The substance has to be carefully analysed and tested to find out what its composition is and what its properties are.

J. Investigate

- Suppose that two new substances have been brought back from a planetary mission. Working together as a group, your job is to decide if they are elements.

- Hold a quick brainstorming session. First, choose one person from your group to write down your ideas. Then, as a group, think of all the different ways you could try to find out if a substance is an element.

- Go through the list carefully and for each idea decide:

  a how it will tell you if it is an element;

  b how you will carry out the experiment to get the answer;

  c how you can confirm your conclusions.

- When you have some clear ideas on how to proceed discuss them with your teacher.

- Organise your group so that you all have clear roles: for example, organiser, experimenter, technician, recorder, data researcher.

- Carry out your investigations on the two substances.

- Make a report of the entire investigation.

METALS AND NON-METALS

Metals are some of the most important materials that we use. Metals were used at least 8000 years ago. By Biblical times, metallurgy was a recognised craft. Tabal-cain was referred to in Genesis as the master of all coppersmiths and blacksmiths. From the time they were first discovered metals have played a major part in civilisation. Two important past eras are now referred to as the Bronze Age and the Iron Age. More than three-quarters of the elements are metals, about 80 in all. Some, like iron, are very common; others, like platinum, are very rare. Metals such as gold and copper have been used since ancient times; others such as titanium and germanium have only become important in recent years. Only about one-quarter of the metals are widely used, but it is hard to imagine life without them. What is it about metals that make them so special?

K. Discuss

• Discuss what you understand by the word 'metal' with your neighbour and arrive at a description you both agree with. Use as many properties as you can. You could write it in the form that you see in quiz books: 'Guess what I am? I am a shiny silver colour. I am ...'

• Now agree on a list of properties for a 'non-metal'.

• When you get home tonight, try out your definitions on two different people. Find out if they work. How will you judge? If necessary, change them and try them out again.

• When you have two good definitions add them to your science word bank.

In the next activity you are going to put your definitions to the test. 'Space blanket' is often sold in camping shops as a survival aid. It is very shiny, just like many metals.

L. Plan and test

• Carry out some tests to find out if 'space blanket' is made from metal.

• Use your findings to suggest why 'space blanket' is advertised as a survival aid. What other uses could it have?

M. Find out   EXTENSION

• The element antimony is described as a semi-metal. Find out all you can about this element.

• Explain why it is called a semi-metal.

A 'space blanket'

The dazzling colour effects of exploding fireworks result from chemical reactions, usually involving metals

The properties of metals and non-metals such as hardness, density, colour and whether or not they conduct heat and electricity are known as their physical properties. You already know quite a lot about the physical properties of both metals and non-metals. You can learn more about metals and non-metals by finding out how they react with other chemical substances. This will give you information about their chemical properties.

N. Investigate

- Using magnesium as an example of a metal and sulphur as an example of a non-metal, design experiments to find out:

  a how they burn;

  b what their oxides are like;

  c how they react with water;

  d how they react with dilute acids.

- Plan your experiments and check them with your teacher before you carry out the experiments.

 Do not stare at burning magnesium. Burn sulphur only in a fume cupboard.

- Make a report of your investigation which clearly shows your results.

- Make a list of differences between the metal and non-metal.

- Look at the periodic table and find the row of elements that begins with sodium, and ends with argon. Find the names of the elements to the left of magnesium and sulphur.

- Predict what the oxides of these two elements will be like.

- Find out if you are correct.

- Suggest a rule for the pH of aqueous solutions of the oxides of metals and non-metals.

- Find out if this rule works for other metals and non-metals.

O. Find out   EXTENSION

- Find out the names and symbols of all the elements in the row of the periodic table that begins with sodium and ends with argon. This row is called the third period.

- Make a chart of these elements across the page in your book.

- Find out the names and formula of the oxides of these elements. Fill them in on your chart.

- Find out the pH of the aqueous solutions of each of the oxides.

- Two of the elements have strange oxides. Which are these elements and what is strange about their oxides?

- Describe the pattern of the oxides across period three of the periodic table.

Elements of the third period			
Name of element	Sodium		
Symbol	Na		
Name of oxide			
Formula			
pH of aqueous solution			

Silicon is used to make many electronic components. It is a hard solid, but it is very brittle. It can be obtained as grey, shiny crystals or as a brown powder. The crystals conduct electricity, but the powder does not. Silicon is said to be essentially non-metallic.

Silicon oxide is a white solid that reacts with alkalis to form salts called silicates. Silicates make up a large percentage of the Earth's crust.

Silicon

P. Think about

- Pretend you *are* silicon. Predict your chemical properties. Find out if your predictions are correct.

- Write two descriptions of yourself, one concentrating on your metal characteristics and one concentrating on your non-metal characteristics.

UNIT 2
FOOD PROCESSES

Section 2.1

FOOD MIXTURES

In many ways our body is like a machine. It takes in materials and by a number of chemical reactions turns them into useful products. In Britain, the food industry provides 70% of the national diet. 50 000 different food and drink products are on sale in our shops. Originally, one of the aims of the food industry was to provide products which had a longer shelf life.

Earlier this century food was produced for the store cupboard

In the 1950s the food industry's main job was to produce food for the store cupboard but recently the need has changed. Many of the food products that are bought now are designed to save time in the kitchen.

Today pre-packed foods make the preparation of meals much less time consuming

Emulsions, gels and foams are important substances in today's food industry. They are all mixtures known as colloids. These mixtures have two components (parts) which normally do not mix. The particles of one component are split up and suspended (dispersed phase) in the other component (continuous phase). In an emulsion both these phases are liquids. A gel is a jelly-like colloid in which a liquid is dispersed in a solid. In a foam, particles of gas are suspended in a liquid.

Some examples of emulsions, gels and foams are given in the table on the left.

Emulsions are very important to how some foods look and taste. Vegetable oils are far too greasy on their own. However, they are valuable ingredients in cooking because they dissolve some flavours that are insoluble in water. If vegetable oils are used as emulsions their greasiness is not so noticeable. French dressing and mayonnaise are both emulsions. They both contain vegetable oils.

Dispersed phase	Continuous phase	Type of colloid	Examples
liquid	liquid	emulsion	milk (oil in water) butter (water in oil)
liquid	solid	gel	jellies
gas	liquid	foam	beaten egg white

vegetable oil
water
egg yolk
spirit vinegar

ORIGINAL
French Dressing

sunflower oil
water
wine vinegar
natural flavours

A. Find out

1 French dressing is a mixture. What does this mean?

2 What are the substances (or ingredients) in French dressing?

3 Are these substances elements, compounds or mixtures?

4 Describe what happens to French dressing if it is shaken vigorously and then allowed to stand.

5 Describe the difference in appearance between mayonnaise and French dressing.

6 Which ingredient is responsible for this difference? Explain your answer.

French dressing and mayonnaise both contain vegetable oil and either vinegar or lemon juice. Mayonnaise also contains egg yolk. The egg yolk contains an emulsifier which helps the oil and the water (from the vinegar or lemon juice) to stay mixed. It increases the attraction between the oil and the water, and prevents them from separating into two layers. Simply, it lets the oil and water 'wet' each other. The diagram shows what happens.

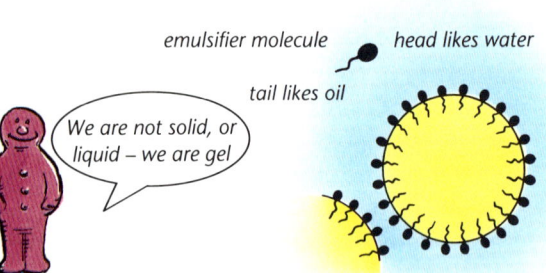

emulsifier molecule

head likes water

tail likes oil

The emulsifier molecules act like a barrier. They stop the oil droplets meeting up to form a layer on top of the vinegar again

water (in vinegar)

The well-known fruit pastille is an example of a gel. It is thought to have originated as a small medicated lozenge that was dispensed by a pharmacist (called an apothecary in days gone by). Over the years the medicinal part has been changed to sugar, and a number of different gelling agents have been used. By using these, a variety of pastilles are now made, ranging from firm wine gums to softer Turkish delight.

We are not solid, or liquid – we are gel

We are mostly water and yet we keep our shape

We have a very special framework

You will see our friends all over the place

B. Investigate

Find out what you can about either emulsions, gels or foams by carrying out *one* of the following investigations. Be prepared to present your findings to the other groups.

1 Emulsions

• Many people prefer not to have raw egg yolk in their mayonnaise. Make some mayonnaise using vinegar and oil, and investigate whether or not soya flour is as good an emulsifier as the egg yolk. Instructions for making mayonnaise are given on Worksheet CPr7.

• Write a detailed report of your investigation. Include in your report a critical evaluation of your methods and results.

• Write a helpful and visually attractive information sheet on 'What is an emulsion?'.

• Look at the labels of different foods to find out which ones contain emulsifiers. (They are in the group of E numbers 400–496).

• Make a display that shows:

  food – emulsifier that it contains – number of emulsifier – (main) source of it.

2 Gels

• Make examples of pastilles using a variety of different gelling agents (starch, pectin, agar). The basic method for making pastilles is shown on Worksheet CPr8.

• When the pastilles are set, test the hardness of each. Try and devise your own simple test for this.

• Work out the cost of making the pastilles. Compare it with the shop prices.

• Write a detailed report of your investigation. Include in your report a critical evaluation of your methods and results.

• Carry out a consumer survey on the pastilles. First you will have to discuss carefully which aspects to include in your survey. (It should not include tasting unless your teacher agrees to this.)

• Write an information sheet on 'What is a gel?'. Make it as helpful and visually attractive as possible.

3 Foams

- Find out how the following affect the quality of the foam made by whisking egg-whites:

  a freshness of egg;   **c** added sugar;

  b temperature;   **d** added egg yolk.

- Make a detailed report of your investigation. Include in your report a critical evaluation of your methods and results.

- Design a simple fact sheet for cooks recommending the best conditions and method of whisking egg-whites (for a stable foam).

- Write an information sheet on 'What is a foam?'. Make it as helpful and visually attractive as possible.

Whisked egg whites are used for making meringues

Whisked egg white is an example of a foam. Air is whisked into the egg-white to make it light. The egg-white is the continuous phase and the air is broken up in this and is the dispersed phase.

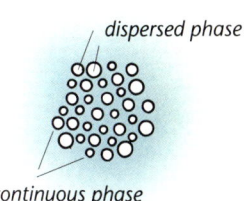

dispersed phase

continuous phase

C. Think about

- Use the information sheets on emulsions, gels and foams to decide which type each of the mixtures in the illustration belongs to.

- Some food colloids need 'additives' to mix together ingredients that would normally separate, and to prevent them from separating again. This group of additives are known as emulsifiers and stabilisers. Thickeners are another kind of food additive. Read the definitions and decide which group of additives each definition belongs to.

> Protect the droplets from colliding and separating out by either thickening the aqueous phase or forming a thin film around each droplet.
> Many obtained from plants.

> Bring oil and water together and mixes them so they do not separate for a short while.
> Many are plant gums and most of them are manufactured from natural products.

> Added to food to increase the viscosity.
> Mostly obtained from plants.

- Using the information you have gained in this section, design a leaflet (such as is given out in supermarkets) with the title 'More about food colloids'. Make the leaflet as interesting as possible. It could contain a puzzle!

EMULSIFIERS

THICKENERS

STABILISERS

COOKING VINEGAR

Some people add vinegar to fish and chips to improve the flavour

In the past, vinegar was far more important than it is today. Pickling was one of the earliest methods of preserving food. Vinegar helps to stop the growth of microbes which cause decay.

Vinegar used to be made at home. Here is a recipe for malt vinegar. The instructions are taken from a cookery book which was published in 1920.

1 Steep the barley in water for several days.

2 Pour off the water and allow the moist grain to germinate. This can be done in the dark where it will imitate the natural conditions of seed germination. You will notice heat being evolved.

3 When the radicle (root) is about half an inch long further germination must be checked by drying.

4 For pale malt dry between 85°F and 95°F; for darker malt dry between 100°F and 120°F.

5 When dry, crush the malted barley and add one pint of water to every ounce.

6 Leave for 2 to 3 hours, shaking occasionally, then filter.

7 Add a little yeast.

8 Leave for a few days.

9 Distil off the alcoholic liquid.

10 Bring the alcoholic liquid into contact with absorbent material which is rich in acetic acid-forming bacteria.

11 Leave for a few days and test the acidity to check the formation of vinegar.

D. Work out

- Making vinegar involves the following chemical processes. Match up the stages to the cookery book instructions.

   a the enzyme diastase turns starch in the barley to maltose sugar; this is called 'malting'

   b alcoholic liquid separates

   c bacteria in old vinegar is involved in a chemical reaction to turn alcohol into ethanoic acid (the acid in vinegar)

   d barley produces the enzyme diastase during germination

   e diastase production stops

   f energy is produced during a chemical reaction

   g malt extract is made

   h pH paper indicates if an acid has been formed

   i yeast 'feeds' on the maltose in the malt extract and makes alcohol and carbon dioxide

- Write an account to show what is happening as the malt is turned into vinegar.

E. Work out   EXTENSION

- The introduction to the recipe for malt vinegar said: *'This recipe makes use of the chief chemical change in the development of a living form from inanimate matter. This is the first step in the production of vegetation which must preceed the animals whose food they compose.'* What did the book mean by each of these sentences?

Making vinegar involves two important biochemical reactions. Biochemical reactions are chemical reactions occuring in living things. Yeast is a very common microbe. It is found on the skins of many fruits and in animal guts. It contains an enzyme (zymase) which can break down certain carbohydrates into alcohol and carbon dioxide. It can do this without oxygen. This chemical reaction is called alcoholic fermentation.

$$\text{maltose} + \text{water} \xrightarrow{\text{malt enzymes}} \text{glucose} \xrightarrow{\text{zymase}} \text{alcohol} + \text{carbon dioxide}$$

$$C_{12}H_{22}O_{11} + H_2O \longrightarrow 2C_6H_{12}O_6 \longrightarrow 4C_2H_5OH + 4CO_2$$

The bacteria then turn this alcohol into ethanoic acid (the acid in vinegar). This process is called oxidation. They do this by adding oxygen to the alcohol.

$$\text{alcohol} + \text{oxygen} \longrightarrow \text{ethanoic acid} + \text{water}$$

$$C_2H_5OH + O_2 \longrightarrow CH_3COOH + H_2O$$

Vinegar contains between 10% and 15% ethanoic acid. It is this which gives it the sour taste.

Two ways of making vinegar are described below. The first, known as quick vinegar making, is a traditional method. Quick vinegar making has been used in European countries where alcohol is free of duty.

Quick vinegar
Alcohol (about 80%) is mixed with 6 parts of water and 1/1000th part yeast. This is warmed to 25°C and poured into the top of the container. From here it trickles down pieces of cord which are attached from the perforated shelf. The liquid runs over wood shavings that have been soaked in vinegar. The holes in the container allow air to enter. As the alcohol is oxidised, the temperature rises to 38°C. The whole process is repeated three or four times and the liquid turns into vinegar in about 36 hours.

The second method uses methanol and is a modern industrial process. According to European Community law, vinegar made by this process cannot be called vinegar! You may see it in the fish and chip shop. It is called non-brewed condiment.

The 'bloom' on ripe fruits such as grapes contains an enzyme that is important in promoting alcoholic fermentation

Stainless steel vessels at a brewery making vinegar by the traditional method

Industrial 'vinegar'

Liquid methanol is reacted with carbon monoxide. To achieve this a temperature of between 180°C and 190°C and a pressure of 30 atmospheres are needed. A catalyst containing rhodium/iodine speeds up the reaction.

$$\text{methanol} + \text{carbon monoxide} \xrightarrow{\text{catalyst}} \text{ethanoic acid}$$

$$CH_3OH + CO \longrightarrow CH_3COOH$$

The reaction provides a high yield (over 99%) of ethanoic acid. However, some of the carbon monoxide reacts with water to form carbon dioxide and hydrogen.

$$CO + H_2O \longrightarrow CO_2 + H_2$$

The advantages of the methanol process are:

1 The methanol can be made from coal, natural gas or oil.

2 The ethanoic acid is the main product.

3 It is relatively easy for the manufacturer to provide the correct conditions for the reactions.

The disadvantage is that the containers for the reaction must be made from special materials. This makes the process quite costly to set up in the first place.

F. Think about

- Compare the two different methods of making vinegar. These questions might help you to get started:

  1 What are the raw materials?

  2 What are the conditions and why are they chosen?

  3 Are there any biochemical reactions? If so, what are they?

  4 What are the products?

  5 What are the advantages and the disadvantages of the two methods?

ALCOHOL AS FUEL

Frequently, food labels give the energy values of their contents. This label comes from a bottle of orange squash.

Nutritional information	A 250 ml serving provides*
Energy	318 kJ / 74 kcal
Protein	0.1 g
Carbohydrate	20 g
Fat	Trace

* diluted 1 part concentrate with 4 parts water

The energy is measured in kilocalories (kcal or Calories) or kilojoules. One kcal is about 4.2 kilojoules (kJ). From the label it looks as though the energy is a separate item and nothing to do with the carbohydrate, fat and protein. This is not so! Virtually all foods provide the body with some energy.

G. Discuss

- Look at each of the food labels in the picture and decide which foods a slimmer would probably choose for inclusion in a Calorie reducing diet.

- What other factors need to be considered when deciding on a Calorie reducing diet?

People do not drink alcohol for the energy content, but alcohol contains a lot of kcal! Half a pint of vintage cider contains about 300 kcal and a gin and tonic about 80 kcal.

Alcohol molecules are small and are quickly absorbed through the gut wall into the blood. The blood then carries the alcohol to all parts of the body. The liver is the organ that removes alcohol from the blood. In the liver, the alcohol is oxidised into carbon dioxide and water and the energy is released. This is very much like the process which occurs all the time in respiration, when sugars are oxidised to water and carbon dioxide and energy is released.

raw mandarin
25 kcal

tinned mandarins
60 kcal

low fat, fruit
yoghurt 135 kcal

boiled carrots
15 kcal

boiled brown rice
35 kcal

strawberry ice
cream 90 kcal

banana
75 kcal

canned lychees
80 kcal

Alcohol is a fuel. Like other fuels, such as oil, coal and gas, it burns in oxygen in a process called combustion to provide energy.

methane + oxygen $\longrightarrow$ water + carbon dioxide + energy

alcohol + oxygen $\longrightarrow$ water + carbon dioxide + energy

sugar + oxygen $\longrightarrow$ water + carbon dioxide + energy

When this combustion occurs in a human cell it is called respiration.

alcohol + oxygen $\longrightarrow$ water + carbon dioxide + energy

sugar + oxygen $\longrightarrow$ water + carbon dioxide + energy

In combustion, or burning, the energy is released as heat. In respiration, the energy is made avilable as biochemical energy for the body to use.

Usually the body does not need to use the energy from alcohol straight away. In this case the alcohol can be turned into fat which acts as an energy store. One word of warning; the liver can only oxidise about 10 cm^3 of alcohol per hour. If the liver regularly receives more alcohol than this it gets damaged. This can result in a condition called cirrhosis of the liver.

Alcohol is sometimes used in cooking

H. Present

- A group of athletes think that drinking alcohol just before a race could provide them with the energy they need. How would you advise them?

The alcohol in drinks is called ethanol. It is just one of a family of alcohols which provide energy when they are burnt.

The names and formulae of some common members of the alcohol family are:

ethanol C_2H_5OH

propan-1-ol C_3H_7OH

butan-1-ol C_4H_9OH

pentan-1-ol $C_5H_{11}OH$

I. Plan and test

- Plan an experiment to find out if there is a pattern between the type of alcohol and the amount of energy it provides. Your apparatus will look something like this:

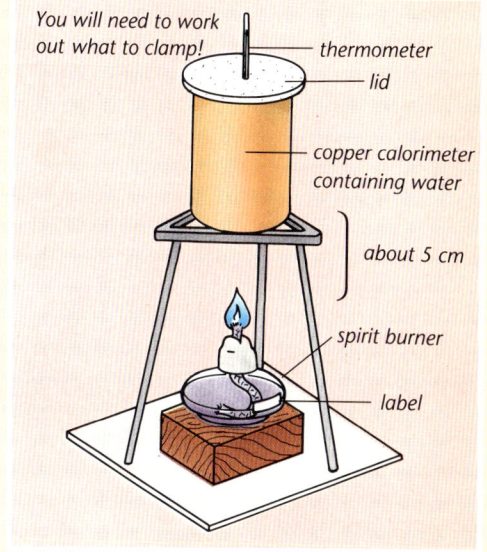

The plan should show:

what you will need to keep the same each time;

what you will need to alter each time;

what you will measure;

how you will present your results.

It should also indicate the safety points you need to keep in mind.

- Carry out the experiment.

⚠ Alcohols are very flammable. They may be poisonous. Keep all bottles and spirit burners stoppered when they are not being used. Keep bottles of alcohol away from flames. Wash your hands after using alcohol and remember to wear your safety spectacles.

- From your results, try and work out a pattern between the number of carbon atoms in the alcohol and the amount of heat given out.

- From your results predict the amount of energy that would be given out by burning 10 g of hexan-1-ol ($C_6H_{13}OH$).

In the experiment you have just done, you found out that chemical reactions can give out energy (in the form of heat) and that the different alcohols gave out different amounts of energy.

You know that another feature of a chemical reaction is that new substances are formed as the reactants are used up. From earlier work in this section you should be able to name the compounds that are formed during the combustion of alcohol.

What compounds are formed when ethanol (methylated spirits) burns?

The atoms in the molecules of a compound are held together by chemical bonds. In this reaction some of the bonds in the alcohol have to break and new ones form. We can think of it like this:

It takes energy to break a bond (just like snapping a ruler takes energy!)

When a bond is formed energy is given out.

The bonds between some atoms are stronger than those between others. In this reaction energy is given out because, on balance, the bonds that are made are stronger than those that have been broken.

The bonds holding atoms in a molecule together can be represented diagrammatically by straight lines. For example, the formula for water can be represented diagrammatically as H-O-H. This shows that there are bonds between each hydrogen atom and the oxygen atom.

The structure of some alcohols

```
  H   H
  |   |
H-C - C - O - H
  |   |
  H   H
```
ethanol

```
  H   H   H
  |   |   |
H - C - C - C - O - H
  |   |   |
  H   H   H
```
propan –1 –ol

```
  H   H   H   H
  |   |   |   |
H - C - C - C - C - O - H
  |   |   |   |
  H   H   H   H
```
butan –1 –ol

J. Work out   **EXTENSION**

- Look again at your results for the previous activity (I).

- Try and explain these results in terms of the number of bonds in the different alcohols.

STOP THE ROT!

You will have noticed that when you are eating an apple, or peeling and slicing apples for cooking, that the flesh of the apple goes brown very quickly. The same thing happens with other fruits, such as pears and peaches. This browning not only spoils the appearance of the fruit but also affects its quality.

This section from a text book for food science students explains why the fruit goes brown.

'Apple goes brown because melanin-type pigments are formed as the plant cells are damaged. Substances called polyphenols, which are usually in the vacuole of the cell are oxidised and turn into hydroxyquinones. This reaction is controlled by the enzyme phenolase which is normally in the cytoplasm of the plant cells. The brown pigments may be unsightly but to the fruit they are important. The products formed during 'browning' are antifungal agents.'

It is the reaction which causes browning in fruits that also causes tea leaves to change from green to their familiar brown colour

---

K. Work out

- Read the explanation of why peeled apples turn brown and then answer the following questions:

  1 How do you know that a chemical reaction has caused browning?

  2 What two conditions change inside an apple as you cut it, to cause this browning?

  3 Try and write a word equation which shows how browning occurs.

  4 Explain how browning is actually advantageous to the apple.

Apples go brown quickly after they have been peeled

Selection of good fruit at the optimum maturity together with careful handling and rapid canning ensure that canned peaches are in the best condition. Heat processing inactivates the enzymes and prevents the oxidation that causes browning

The spoiling of fruit by browning is very important in the large scale commercial production of fruit products. It affects the quality of apple juice, apple pulp, canned peaches and pears, dried fruit, cider and other products. In the next activity you are going to investigate ways of solving the food browning problem. It is important that the remedy does not spoil the flavour or texture of the fruit or use toxic materials.

L. Investigate

- Working in a group, first consider the factors that you think may cause fruit to go brown.

- Then discuss the ways in which you may be able to prevent browning.

  Do any of your family have ways of stopping fruit or potatoes going brown when they have been peeled?

  How do they stop potatoes going brown in chip shops?

- Make a list of all your ideas.

- Turn your ideas into a plan of action. Think about what you are trying to find out, what you must change, what you must keep the same and how you will record your results. Remember, you are dealing with a foodstuff, so all the things that you try must be 'safe'.

- Check the plan carefully and then carry out the investigation.

- When you have finished (and cleared away!) prepare a report, including some simple practical steps that a restaurant could take to avoid browning.

Additives are frequently added to processed food to prevent oxidation occurring. Fats and oils become oxidised when they are left in the air. When this happens the food containing the fat goes rancid and tastes 'off'. This can make you sick if you eat the food. Antioxidants are added to stop this, as well as to stop fruit browning.

Traces of metals such as copper, iron and nickel may occasionally get into the food from the machinery used in the processing. These can accelerate the oxidation of food. Substances, called sequestants, are sometimes added to food to help to prevent this. These attach themselves to the metals and so stop them speeding up the unwanted process of oxidation.

M. Find out

- Look at the labels of several foods to find out which ones contain antioxidants and sequestants.

- Make a display that shows:

  food – anti-oxidant or sequestant – number – (main) source of it.

UNIT 3
CONTROLLED REACTIONS

Section 3.1

A SHARED LANGUAGE

The pupils in a school in France are studying the topic 'Controlling chemical reactions'. The picture sequence shows part of one of their lessons.

1 Observez ce qui se passe. Est-ce qu'il y a toujours une réaction?

$$2H_2O_2(aq) \rightarrow 2H_2O(\ell) + O_2(g)$$

H₂O₂ | H₂O₂ | H₂O₂ — 1 2 3 — SABLE MnO₂ FOIE

2 D'abord j'ajoute le sable ... euh non, rein!

H₂O₂ — SABLE — 1

3 Ensuite la meme quantité d'oxyde de manganese(IV) á la deuxieme et de la foie á la dernière.

H₂O₂ — MnO₂ — 2

4 Maintenant répondez à ces questions.

1. Quelle est la réaction?
2. Quel est l'agent réactif?
3. Quels en sont les produits?
4. Dans l'équation, que signifie chaque symbole qui indique la forme de la matiere?
5. Quelle réaction est la plus rapide?
6. L'oxyde de manganese (IV) et la foie accélérant la réaction. On les appelle des "catalyseurs". La foie contient des "catalyseurs" vivants. Comment s'appellant-ils?

H₂O₂ | H₂O₂ | H₂O₂ — 1 2 3

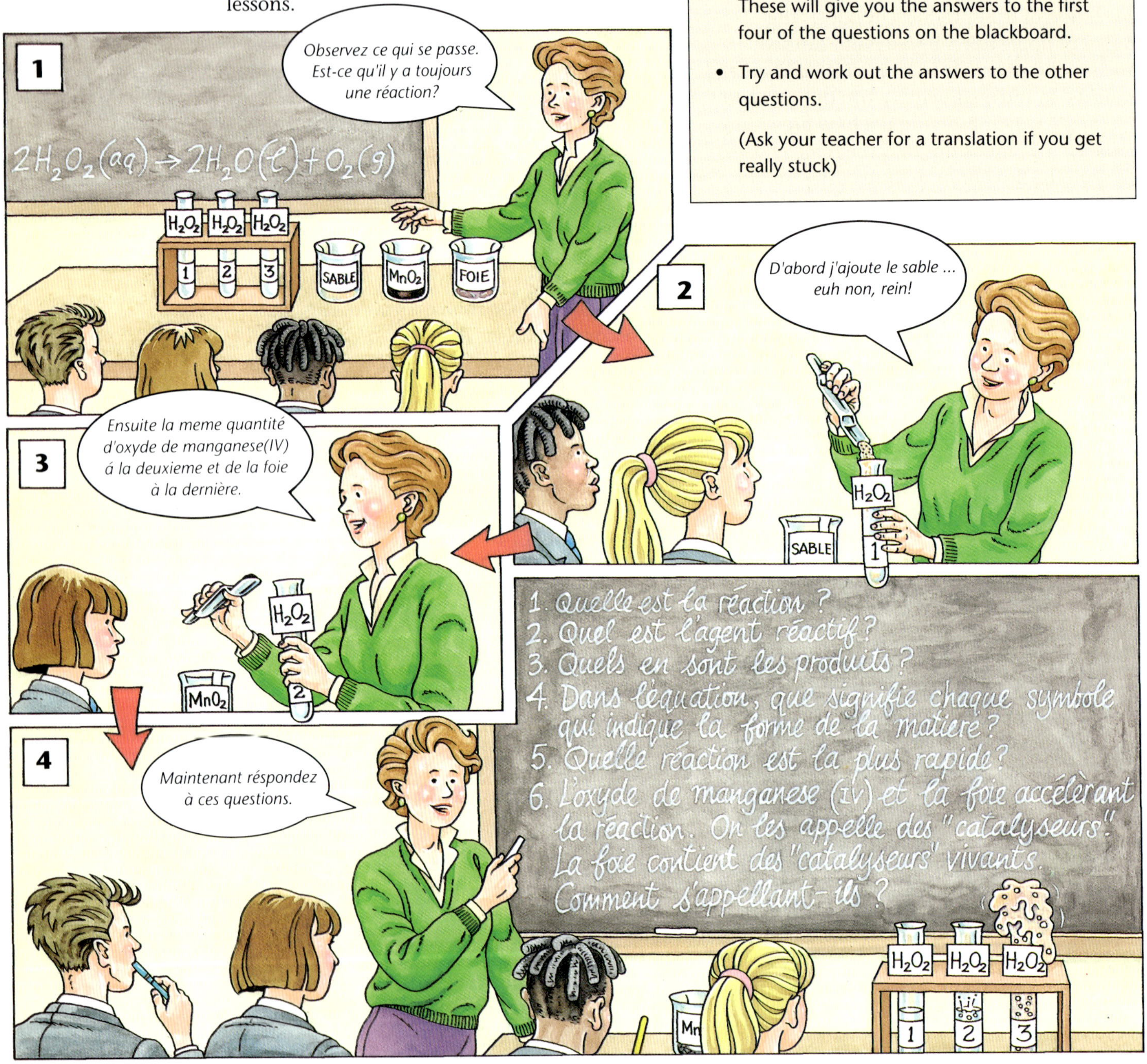

By using symbols for chemical elements people who normally speak in different languages can 'talk' to each other about chemical reactions. The element we call oxygen in English is known as oxygène in French, Sauerstoff in German, oxígeno in Spanish and ufelai in Welsh. The symbol for oxygen is O. This symbol is used throughout the world. Whenever it is used a scientist will understand what it means.

It was a Swedish chemist, Jons Berzelius, who first saw how important the use of chemical shorthand would be. He spoke little English. Perhaps it was because of his need to communicate with other famous chemists of the time, like Humphry Davy in England, that he gave symbols to the forty elements that were known then.

If you look back to the pictures of the French lesson you will see the formulae of two compounds that contain the elements hydrogen and oxygen written on the board. These two compounds are very different substances, with very different properties. It is important to know the difference!

You already know that you can write water as H_2O and that this means that water is made up of the two elements hydrogen (H) and oxygen (O). We are now going to find out why water is written as H_2O and not as OH, H_2O_2 or HO_2. First we need to look back at how formulae for compounds came to be written.

The first important step was made by Robert Boyle in the 1670s. He proved that 'he could create nothing to take the place of that which he would destroy'. Later, in 1774, Antoine Lavoisier experimented with heating elements in a closed container. He found that there was no change in mass. This led to the development of the law of conservation of mass, which says 'if we measure the total mass of the reactants and allow the reaction to happen and then measure the total mass of the products we will not find a change'. In other words, atoms can neither be created nor destroyed during the course of a chemical reaction.

B. Plan and test

- Copper(II) sulphate solution reacts with sodium hydroxide solution to form copper(II) hydroxide (a solid) and sodium sulphate (solution). Use this information to plan (and carry out) an experiment to show that the law of conservation of mass is really true for this reaction.

- Write a word equation for your reaction.

- Explain what has happened to the mass of the copper during the reaction.

- Explain what has happened to the bonds in the compounds during the reaction.

Landolt developed a special H-shaped piece of apparatus known as the Landolt tube. In 1906, using this piece of apparatus, he investigated 15 different chemical reactions. He placed the reactants in the two 'arms' of the tube and sealed the tube by heating before weighing it. He then allowed the chemicals to mix. After the reaction was complete he reweighed the tube. It is recorded that he found that the change in mass during the reaction did not exceed 1 part in 10 000 000!

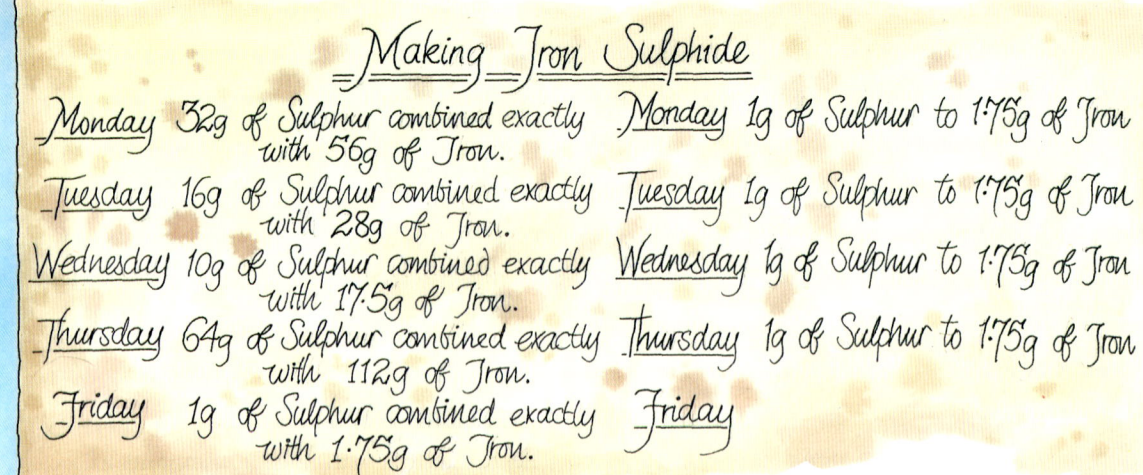

Making Iron Sulphide

Monday 32g of Sulphur combined exactly with 56g of Iron.
Monday 1g of Sulphur to 1·75g of Iron

Tuesday 16g of Sulphur combined exactly with 28g of Iron.
Tuesday 1g of Sulphur to 1·75g of Iron

Wednesday 10g of Sulphur combined exactly with 17·5g of Iron.
Wednesday 1g of Sulphur to 1·75g of Iron

Thursday 64g of Sulphur combined exactly with 112g of Iron.
Thursday 1g of Sulphur to 1·75g of Iron

Friday 1g of Sulphur combined exactly with 1·75g of Iron.
Friday

An extract from the notebook of a chemist experimenting at the beginning of the nineteenth century

The law of conservation of mass enables chemists to work out how much of each reactant is needed for a chemical reaction and how much of each product will be made. This is often very important. For example, chemical engineers who manage industrial chemical processes need to know how much product will be available to sell.

Why do chemical engineers need to know how much product will be available to sell?

Knowing that the total mass stays the same during a reaction does not help us with knowing why water is always H_2O and not sometimes HO, HO_2 or H_2O_2.

During the eighteenth century many chemists were beginning to believe that each compound had a fixed composition. They thought that in a compound the elements were combined in certain fixed proportions.

The results shown in the notebook at the top of the page indicate that whenever the compound iron sulphide is prepared it always contains its elements in fixed proportions. Chemists tried hard to understand how, and in what proportions, elements combined together. They used phrases such as 'combining weights', 'reacting quantities' and 'the law of equivalents'.

The real breakthrough was made by John Dalton, in 1808. From the results of experiments he deduced that when elements combine to form compounds the atoms of these elements combine in small whole numbers. It was his work that led to the use of formulae to give information about compounds.

The formula of a compound tells us what elements are present in the compound and the proportions in which the atoms of the elements are combined. The formula for water, H_2O, tells us that two atoms of hydrogen are combined with one atom of oxygen in each water molecule. Formulae are also used in equations to describe chemical reactions.

C. Work out

- Fill in the missing amounts for each of these reactions:

1   sulphur + oxygen ⟶ sulphur dioxide
    32 g   32 g   ?

2   carbon + oxygen ⟶ carbon dioxide
    12 g   ?   44 g

3   carbon + oxygen ⟶ carbon monoxide
    12 g   ?   28 g

4   hydrogen + oxygen ⟶ water
    2 g   16 g   ?

5   hydrogen peroxide ⟶ water + oxygen
        34 g   ? + 16 g

So far we have learnt these two facts about chemical reactions:

A No atoms are lost during a reaction. All the atoms that are there to start with in the reactants must be there at the end, in the products.

B Elements combine to make compounds in simple proportions.

We can use these facts to help us to write equations.

The steps for writing an equation are as follows:

1 Write the description of the reaction in words.

2 Replace 'goes to' with $\longrightarrow$, and 'and' with +.

3 Replace the names of the substances with their formulae. (The most common formulae you come across are given in Worksheet CPr30.)

4 Count the number of atoms of each type on each side of the equation.

5 Balance the equation.

6 Check that the number of atoms of each type is the same on both sides of the equation.

7 Finally, add the correct state symbols: (g) for gas, (l) for liquid, (s) for solid, and (aq) for dissolved in water.

---

1 hydrogen peroxide goes to make water and oxygen

2 hydrogen peroxide $\longrightarrow$ water + oxygen

3 $H_2O_2 \longrightarrow H_2O + O_2$

4

	Reactants	Products
hydrogen	2	2
oxygen	2	3

It looks as if some oxygen has been made. This can't be true.

5 The equation cannot be balanced by changing the formula of any of the substances. The equation would balance if H_2O_2 was changed to H_2O_3, but this would be an entirely different substance. It is only possible to put numbers in front of the substances. This just changes the amount of substance used.

$2H_2O_2 \longrightarrow 2H_2O + O_2$

6

	Reactants	Products
hydrogen	4	4
oxygen	4	4

So this is the balanced equation for the reaction.

7 $2H_2O_2(aq) \longrightarrow 2H_2O (l) + O_2(g)$

This final equation tells us that two molecules of hydrogen peroxide decompose to form two molecules of water and one molecule of oxygen. It also tells us that two moles of hydrogen peroxide molecules decomposed to form two moles of water molecules and one mole of oxygen molecules.

A mole is the word used in chemistry for a particular amount of substance.
Worksheet CPr12 gives you more practice at balancing equations and explains more about moles.

GRAVE REACTIONS AND PAIN KILLERS

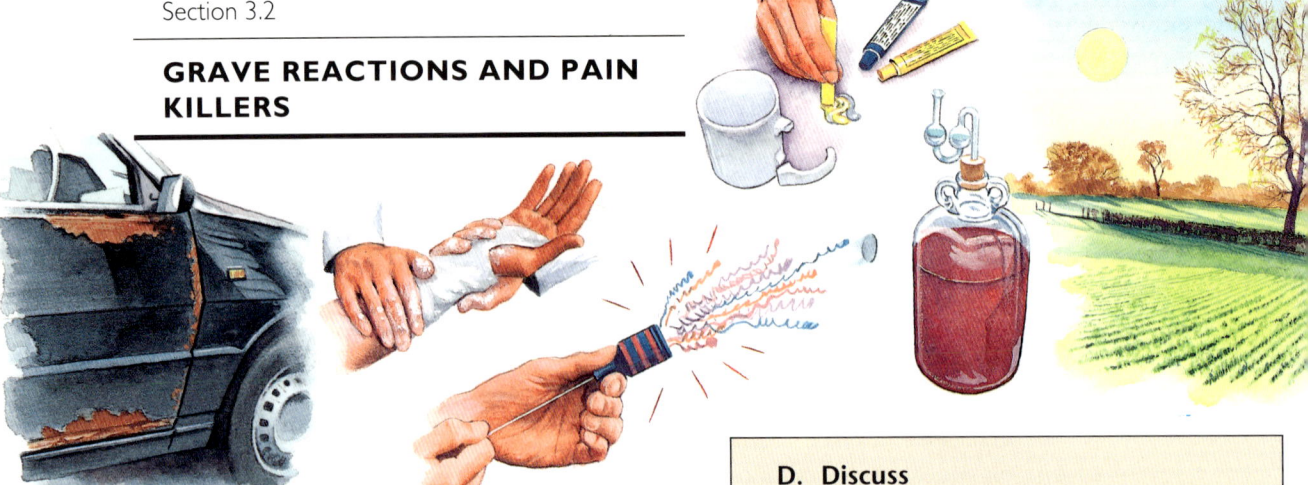

A few chemical reactions that you might see today

To compare reaction speeds fairly they must be measured in some way. The speed (we usually call it the rate) of the reaction can be measured either by measuring how fast the reactants get used up or by measuring how fast new products are made. Another way of saying this is:

$$\text{rate of reaction} = \frac{\text{amount of new substance formed or used up}}{\text{time}}$$

D. Discuss

- How many chemical reactions can you spot in the illustration?

- Try and put the chemical reactions you have spotted in order of their speed. Call the fastest reaction number 1.

  Why is it difficult to decide on the speed of the reaction?

E. Investigate

Look at all that detail. That's been here since about 1988

But look at this one. You can hardly make out the date, it's 1870. What is it in the air that's made it change like this? I think they're made of limestone aren't they?

Must be the fumes in the air from that factory. They pump acid into the air you know. I'm sure it's got worse recently – they must be using stronger acid

I don't think we'll be able to read this in a few years it's all powdery and the atmosphere will get at it more. I wonder if it deteriorates quicker in the summer or the winter?

- Read the picture story about Ben and Claudine.

  What reaction do Ben and Claudine suspect is going on at the gravestones?

  Which factors do they think are affecting the rate of this reaction?

- Investigate this gravestone reaction and the factors that affect it.

  Begin by planning your whole investigation. You will need a detailed list of materials and equipment and a written 'plan of action'. Think about what data you will need to collect. You will probably need to carry out a few trial experiments to decide on sensible amounts of substances to use.

  Check your plan with your teacher. If the plan is alright, carry out your investigation.

- After you have finished the investigation (and the clearing up), prepare a report.

Reaction rates are important in our everyday lives. When we take medication for a headache we want it to act as quickly as possible. An analgesic (pain killer), product number 0821, is being investigated by the research and development department of a pharmaceutical company. The investigation concerns whether the product should be sold as a tablet or in powder form. When added to water, the product dissolves and is more readily absorbed into the bloodstream. The rate of gas production gives an indication of how fast the product will dissolve. Some results of the investigation are shown in the tables.

Tablet version: 0821T

0.5 g of the tablet added to 30 cm³ of water

Time(s)	0	30	60	90	120	150	180
Volume of gas (cm³)	0	16	30	35	47	51	51

Powder version: 0821P

0.5 g of the powdered version (0821P) was added to the same volume of water

Time(s)	0	30	60	90	120	150	180
Volume of gas (cm³)	0	37	43	50	51	51	51

F. Work out EXTENSION

- Use the results given in the tables to plot a graph of the rate of gas production for each investigation. Put time on the horizontal axis.

  1 Which reading in the tablet investigation looks incorrect?

  2 If you had a bad headache, which of the product versions would act fastest?

  3 What other factors are there to consider when deciding to market this version of the drug? (You can assume that it has passed trials to ensure its safety.)

G. Plan and present

- 'Changing the rate of chemical reactions' is the title of a short talk which is to be illustrated by a set of six slides. These slides must show a good range of the practical problems taken from everyday life and industry. Each slide should be accompanied by a few simple notes about the need to change the rate of the reaction. Your task is to decide what the six slides should show, prepare the design briefs for the slides and write the accompanying notes. (It may help to sketch the slide, or to use a photograph from a magazine and say how this should be changed.) An example is done for you.

Design brief for slide
Person taking meat out of shopping basket and placing it in a freezer.

Accompanying notes
Meat deteriorates quickly at room temperature. Freezing preserves food because it slows down the chemical reactions that cause the deterioration.

Water being sprayed on a coal face as it is being cut. The water causes the particles of dust to clump together. How could this help to prevent an explosion taking place?

CHANGE WITHOUT CHANGING

By now you should be becoming quite experienced with chemical reactions. You should be able to:

1 Write an equation that shows the reactants and products.

2 Suggest how to measure the rate of a reaction.

3 Suggest some ways of speeding up or slowing down a reaction.

Some reactions can be speeded up by substances called catalysts. A catalyst is a substance that speeds up a chemical reaction but is not used up by the reaction.

Hannah and Robert carried out an experiment to find out how adding different substances to hydrogen peroxide affected how fast it decomposed. The aim of the experiment was to find out what was the best catalyst for this reaction. They included this graph of their results in the report of their experiment:

A selection of catalysts

Graph to show the decomposition of hydrogen peroxide

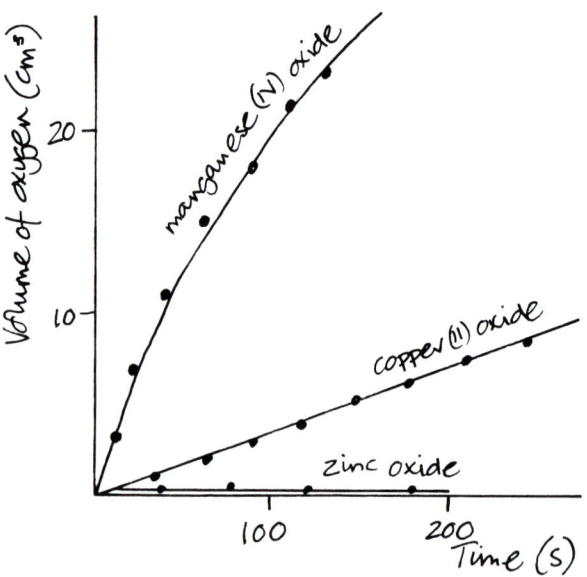

H. Work out

- Study the graph carefully and then write a suitable conclusion for the experiment.

Catalysts are very important substances and they are widely used by the chemical industry.

I. Work out   **EXTENSION**

- Read through the following information about catalysts and work out how each stage contributed to our understanding of catalysts.

- Present your report in an imaginative way.

The following are steps that have led to our present understanding of catalysts:

1 In ancient times people knew that very small amounts of certain substances were needed for fermentation to occur.

2 Alchemists knew of the great effects of small quantities of some substances; they called these substances 'quintessences'.

Alchemists knew of the great effects of small quantities of substances which they called 'quintessences'

3 Mrs Fulhame, 1794, found out that 'to oxidize carbon monoxide needs a small quantity of water.'

4 Gottlieb Kirchoff, 1811, discovered that 'when acid turns starch to sugar it is itself unchanged at the end.'

5 Wolfgang Dobereiner, 1822, observed that 'oxygen and hydrogen will combine at room temperature, if there is a little finely-divided platinum present.'

6 Dubrunfaut, 1830, discovered that '*malt extract works in the same way as an acid when present with starch.*'

7 Henry Phillips, 1831, patented his process; '*in the presence of platinum, sulphur dioxide readily combines with oxygen to form sulphur trioxide. When this dissolves in water sulphuric acid is formed. The platinum is not consumed in the reaction.*'

8 Jons Jacob Berzelius, 1835, used the word '*catalysis*' to describe one substance, by its presence, causing others to react.

9 Alexander Williamson, 1854, became the first person to explain how a catalyst worked.

10 Buchner and Haln, 1897, found out that cell-free extracts produced the same effect as the organisms that the cells came from.

11 Friedrich Ostwald, 1888, discovered that catalysts and enzymes do not start reactions, or alter the actual products, they only affect the reaction speed. He later received a Nobel prize for his work on catalysts.

12 Catalysis led to the development of the Contact process for sulphuric acid production – sulphuric acid is often called the most important of all chemical substances. Catalysts are also used to make nitric acid using nitrogen from the air. Nitric acid is used in the fertiliser and explosives industries. Until nitrogen catalysts were developed, Germany had to rely on imported nitrates from Chile as their source of fertilisers. Some historians think that it was the ability to manufacture explosives using catalysts that enabled Germany to go to war in 1914 – 1918.

The human body contains thousands of enzymes. Enzymes have been used for centuries in food production. Zymase, present in yeast, is used to turn sugars into alcohol during fermentation. More recently scientists have started to use enzymes to bring about other useful reactions. Many washing powders contain enzymes called proteases. These act on the proteins in stains like blood and egg.

Other enzymes are important in medicine. An enzyme, chymopapain, is used to tenderise meat. This same enzyme can help some people who have a slipped disc to avoid an operation. The enzyme can be injected into the area of the slipped disc. Here it acts on the material causing the bulging and so reduces the pressure on the nerve which causes pain. Cortisone is a drug used to treat arthritis. It used to cost £100 per gram. Now, by using enzymes in its manufacture, the cost has been reduced to less than 50p per gram.

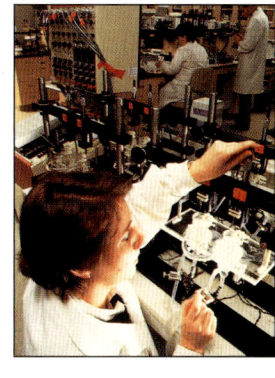

Chemists working to develop drugs to treat the symptons of arthritis and rhuematic diseases

J. Research

- Find out what the difference is between an enzyme and a catalyst.

- Find out about the use of catalysts and enzymes in some of the following processes:

  production of margarine

  production of nitric acid

  production of ammonia

  production of sulphuric acid

  enzymes in digestion and biotechnology

  catalytic converters

- Record your findings in a table similar to the one shown below.

Process	Reactants	Products	Name of Catalyst or enzyme	Notes
Decomposition of hydrogen peroxide	hydrogen peroxide	water and oxygen	manganese (IV) oxide enzyme in liver	can occur in cells of body

UNIT 4
THE SALT INDUSTRY

Section 4.1

MANUFACTURING INDUSTRIES

Bus stations, railway stations and airports are places where people are often forced to pass the time. Imagine that you have a delay of an hour in your journey! You could pass your time analysing the manufactured items around you. Most of the raw materials which are used in manufacturing goods come from the air, water, rocks, living things and fossil fuels.

A. Think about

• Imagine that you have to spend some time in the airport lounge shown in the picture. Look around you at all the items that have been manufactured. Decide what substances each item has been made from, and from what raw material each substance came.

• Spend ten minutes filling in a table similar to the one shown here. Try and make the information as comprehensive as you can. One example has been done for you.

Item	Substance	Raw material
red nylon jacket	nylon	oil
	dye	probably oil or coal

• Look through your list of raw materials. Does it contain examples from each of the sources listed? If not, look at the picture again and find at least one substance made from a raw material from each source. Add these to your table.

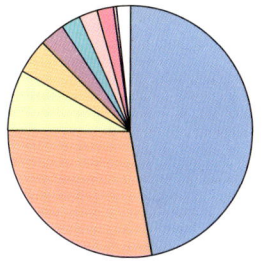

oxygen 47%
silicon 28%
aluminium 7.8%
iron 4.5%
calcium 3.5%
potassium 2.5%
sodium 2.5%
magnesium 2%
titanium 0.5%
others 1.7%

Percentage by mass of elements in Earth's crust

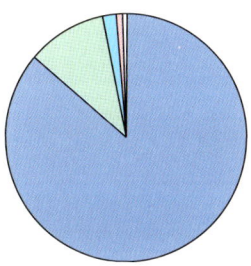

oxygen 86%
hydrogen 10.9%
chlorine 1.8%
sodium 1%
others 0.3%

Percentage by mass of elements in the sea

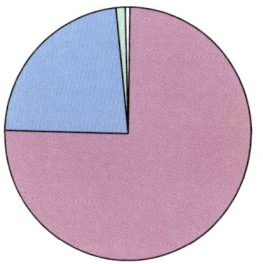

nitrogen 75.5%
oxygen 23%
argon 1.4%
others 0.1%

Percentage by mass of elements in the atmosphere

Turning raw materials into useful products occurs during manufacturing. The word 'manufacture' simply means to make something from a raw material, although most people think of a manufacturing process as a large scale operation using machinery.

B. Research

• Find out about the manufacturing industries near you.

  What do they make?

  What are their raw materials?

• Explain what the term 'a manufacturing industry' means to you.

Raw materials come from the air, water, rocks, living things and fossil fuels; but how did they get there? Boyle believed that metals and minerals 'grew' in the earth. He wrote *'we may deduce that earth by a metalline plastick principle latent in it ('seed') may in process of time (be) changed into a metal.'* Many scientists at the time agreed with him.

The three components of our planet Earth – the solid crust, the water (covering 70% of the surface) and the atmosphere, provide the elements which give us our raw materials and enable us to make new materials.

The raw materials are not spread evenly throughout the Earth's crust. They tend to be concentrated together in areas. Major chemical industries are sited near concentrated sources of raw materials. Tracing the development of an industry can tell us much about the development of chemical processes.

The development of the textile industry had a profound effect on the industrial development of this country. During the early eighteenth century several major advances in agriculture occurred. This agricultural revolution meant that the quality of life improved above subsistence level, and the population began growing rapidly. As this happened, there was a greater demand for cloth, especially cotton. Up until this time most goods had been made by hand in people's homes and often the whole family were involved. Some of the first changes brought about by the increased demand for cloth involved the use of more specialised equipment at home, resulting in framework knitters and weavers using more complicated looms. These were soon replaced by factories where the machines could be power driven.

The interior of a cotton mill in 1842

Mechanical cutter extracting salt in the Winsford salt mine, Cheshire

The main textile producing areas in 1830

The main chemical industry areas today

The factories were built as close as possible to the sources of the raw materials. Cotton from America and Brazil arrived at the port of Liverpool. The area around Liverpool already had a plentiful supply of soft water, used at first for power and later in the important processing and cleaning stages. The damp Lancashire climate was ideal for cotton spinning. Later coal was used to provide the power, and this too was available locally. Many subsidiary industries started to develop – iron (for the machines), alkalis (for treating the woven cotton), soap, dyes and mordants were all needed.

Until this time chemists had worked in isolation, often carrying out experiments just because they were interested in them. Now they were much in demand by industry. They were employed by factory owners to help to make the production of the materials more profitable.

The alkali industry played a big part in the success of the developing textile industry. It still has a large effect on our lives today. The substance that is at the centre of this industry is common salt (sodium chloride). One of the few working salt mines in Britain today is at Winsford in Cheshire. It has been in existence since 1844 and at present has about 24 million tonnes of recoverable rock salt. About two million tonnes of this are extracted each year. Most of it ends up being spread on icy roads. To be used in industry the salt needs to be pure.

In the following activities you will be able to make your own alkali from rock salt such as might be mined at Winsford. The processes will be similar to the main processes in the chlor-alkali industry.

C. Plan and test

- Your problem is to obtain pure salt from a sample of rock salt and to work out what percentage (by weight) of pure salt there is in the sample. Plan carefully what to do. You will need to work out how you are going to find the percentage of pure salt before you begin.

- Carry out your purifying process. Keep the purified sample to use in Section 4.2.

- When you have finished draw a series of diagrams to describe your method.

  How much salt did your sample contain?

  How much waste was there?

PROCESSING SALT

In this section you are going to look at some of the important processes in the chlor-alkali industry. You will need the sample of purified salt which you prepared in the last section, to make into a solution of a certain concentration. The labels on bottles of chemicals often show concentrations, for example, 1M H_2SO_4 or 0.5M $CuSO_4$(aq). The number in front of the M tells you how concentrated the solution is. You are going to make a 1M solution of sodium chloride. This is more correctly called a concentration of 1 mol dm^{-3}. The solution contains the molar mass of sodium chloride made up to a decimetre cubed (litre) of water. This concentration works well for your chlor-alkali process. The molar mass of sodium chloride is 58.5 g. To make a 1 mol dm^{-3} solution, distilled water is added to make this up to 1000 cm^3 (1 dm^3). You will only need 100 cm^3 of solution, so you will only need 5.85 g of sodium chloride.

The mass of one mole of substance is called the molar mass.

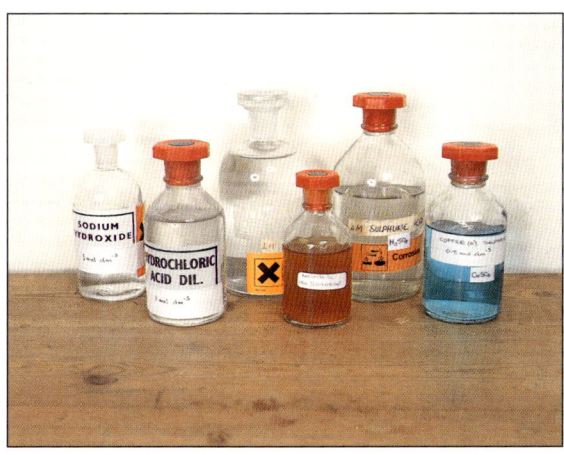

Bottles of chemicals showing their concentrations

D. Work out

- Weigh out 5.85 g of sodium chloride (your purified salt) and put it in a clean 100 cm^3 beaker.

- Add distilled water (no more than 50 $cm^{3)}$ and stir until the sodium chloride has dissolved.

- Gently pour the salt solution into a volumetric flask – you must use a filter funnel.

- Rinse the beaker well with distilled water, making sure all the liquid goes into the volumetric flask.

- Add distilled water until the level is about 1 cm^3 below the mark on the neck of the flask. Insert the stopper and swirl the contents of the flask gently to mix.

- Now, using a dropping pipette, add distilled water drop by drop until the bottom of the meniscus just rests on the mark – make sure your eye is level with the meniscus when you check this.

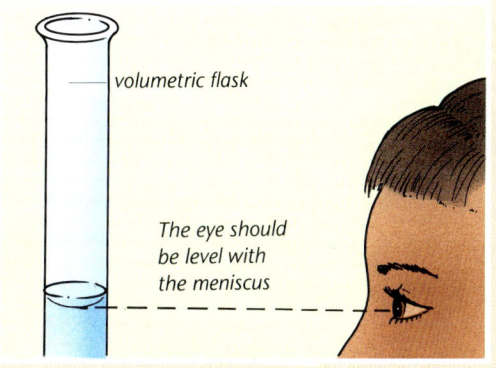

volumetric flask

The eye should be level with the meniscus

- Insert the stopper and shake the flask thoroughly to mix the contents.

- Label your volumetric flask. You will need this solution for your next activity.

How would you make 100 cm^3 of:

a   0.5 mol dm^{-3} sodium chloride;

b   2 mol dm^{-3} sodium chloride?

You are now ready to carry out the electrolysis of your sodium chloride. The word electrolysis means 'splitting by electricity'; in other words, causing a chemical reaction in which a compound is split up, in this case sodium chloride. This is the central process of the chlor-alkali industry.

E. Observe and record

- Set up the apparatus shown in the diagram in a fume cupboard.

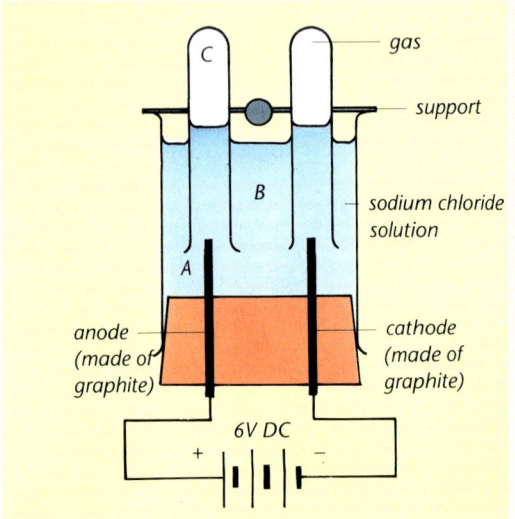

What two compounds are there in the cell?

- Test the pH of your solution.

- Switch on the current and watch the electrodes carefully.

- Predict what gases are being formed at each electrode.

- Use indicator paper to test the pH of the solution at A and B and the gas shown at C in the diagram.

 Be careful not to put your hands in the solution.

- Collect a sample of the gas coming from the cathode and test your prediction.

- Draw a diagram of your apparatus and annotate it to show all that you noticed during the experiment.

Sodium chloride is an ionic substance. This means that when it is dissolved in water, the liquid conducts electricity (the liquid is called an electrolyte) and chemical changes occur at the electrodes. In the solution there are two compounds, sodium chloride(NaCl) and water (H_2O). This means that in sodium chloride solution there are four elements involved: sodium, hydrogen, chlorine and oxygen.

The following rule will help you to work out and remember what happens at each electrode:

> **Metals and hydrogen appear at the cathode**
> **Non-metals appear at the anode**

This rule implies that sodium and hydrogen will appear at the cathode and oxygen and chlorine at the anode. You saw gases at both electrodes and should have correctly identified them as hydrogen at the cathode and chlorine at the anode.

Another rule to help you to predict what will happen during electrolysis is this:

> **At the cathode, hydrogen appears in preference to metals**

(The only exceptions you are likely to meet are copper, nickel and other unreactive metals)

> **At the anode, chlorine, bromine and iodine appear in preference to oxygen**

The overall reaction for the electrolysis of sodium chloride solution is:

sodium chloride + water $\longrightarrow$ sodium hydroxide + chlorine + hydrogen

$$2NaCl + 2H_2O \longrightarrow 2NaOH + Cl_2 + H_2$$

F. Work out

- If the two gases chlorine and hydrogen are given off during the electrolysis of sodium chloride solution, work out what is left in the electrolysis cell.

- Write the two rules for helping you to work out what happens at the electrodes during electrolysis under your electrolysis diagram.

Part of a plant producing chlorine by the electrolysis of brine. The pipes are colour-coded for safety:
yellow = chlorine
red = hydrogen
green = water

You should have noticed that when you placed a pH paper at A in your electrolysis cell it was bleached white. The chlorine reacts with the alkaline solution to give sodium chlorate(I) which is sodium hypochlorite bleach.

sodium hydroxide + chlorine $\longrightarrow$
sodium chloride + sodium chlorate(I) + water

$$2NaOH + Cl_2 \longrightarrow NaCl + NaOCl + H_2O$$

G. Work out

- The production of sodium chlorate(I) is a big problem to the chlor-alkali industry. Try and work out what could be done to prevent this happening.

Electrolysis is taken very much for granted these days. In fact, electricity is also taken very much for granted. Just think for a moment what today would have been like for you if you had not been able to rely on electricity.

In the eighteenth century all the electrical experiments were carried out using what was called frictional electricity. This is now called static electricity. In 1786 Luigi Galvani carried out an experiment which was to become very famous. He discovered that frog's legs, even

when they were not attached to the frog, would twitch if they were connected to a length of copper and iron wire.

Galvani was convinced that electricity was produced, in some way, by the muscles and nerves. Another Italian, Volta, thought that the two different metals were more important than the frog's legs in producing this electricity. He concentrated on producing electricity when two different metals were separated by non-animal liquids. In 1800, he announced his invention. It was known as the Voltaic pile and was the first source of a steady electric current. In an early Voltaic pile Volta used silver and zinc discs separated by cardboard soaked in brine.

Other researchers were quick to use this invention. One of the first uses was for electrolysis. In 1800, Nicholson and Carlisle discovered that they could split up water, a compound, into its elements by using an electric current. Many compounds were known at the time. Chemists knew that compounds such as silver nitrate, copper(II) sulphate, sodium chloride and sodium nitrate could be formed from acids and bases and that bases were oxides of metals. However, nobody had managed to extract the metals from these compounds. Common salt was especially interesting as it contained sodium. Humphry Davy passed a current through a solution of the salt but he failed to obtain the metal. At the same time as he was conducting his experiments, better batteries were being invented.

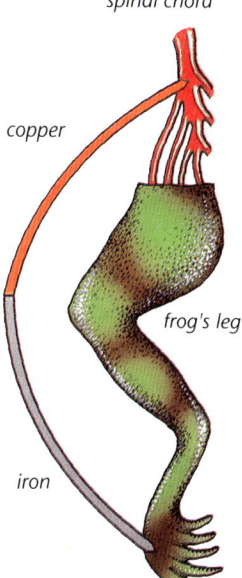

spinal chord

copper

frog's leg

iron

One of Humphry Davy's early batteries

In 1801, using batteries containing thousands of cells, Davy passed electricity through sodium hydroxide and potassium hydroxide. He produced sodium and potassium and so confirmed that these alkalis were metal-containing compounds. During the same year he proceeded to isolate other elements – barium, strontium and calcium in the same way. He also predicted the existence of magnesium but this was not isolated until 1829, 21 years later.

H. Research

- Use the information in this section, and any other information that you can find, to make a poster about the discovery of electrolysis and how it was first used.

Davy was an excellent lecturer who attracted large audiences. He frequently lectured at the Royal institution in London, which was established in 1799 to show how teaching and research could be applied to everyday life. Having heard Humphry Davy lecture at the Royal institution and being eager to 'enter into the service of science' the young Michael Faraday applied to Davy for employment. As a result he was, at the age of 21, appointed assistant to Davy to help with both lectures and experiments. The two worked together for many years.

Michael Faraday was another researcher who was interested, like Davy, in electricity and chemistry. In the early 1830s, as a result of his researches, Faraday worked out the basic rules of electrolysis that we still use today.

Faraday's work also provided a clue to the existence of the electron as the 'unit' of electricity.

THE CHLOR-ALKALI INDUSTRY

The chlor-alkali industry is the part of the heavy chemical industry which manufactures chlorine and sodium hydroxide (caustic soda). The chemical industry is frequently divided into two parts – one part dealing with heavy chemicals and the other with fine chemicals. There is no exact definition of a heavy chemical; the term refers to those chemicals that are used in large quantities, such as sodium hydroxide, sulphuric acid and lime. Fine chemicals are those that tend to be costly and used in small amounts. They are of high purity and are chemically complex. They include chemicals such as drugs, vitamins and saccharin.

Chlorine and sodium hydroxide are now almost entirely made by the electrolysis of salt. Both chlorine and sodium hydroxide are intermediates in the manufacture of a large number of other substances. You will find out about these in Section 4.4.

In 1830 the chemical industry was based around the textile industry and involved mainly acids, alkalis, mordants, dyes and those other chemical substances needed by the textile industry. At the beginning of the Industrial Revolution (about 1750) treating the woven cotton cloth (called finishing) was a costly, primitive and slow job. Bleaching alone took eight months! Research laboratories were set up in finishing works and, although many of the details of such work were well guarded secrets, meetings were held to exchange general scientific information. It was at one such meeting that John Mercer, in 1850, described the action of caustic soda (sodium hydroxide) on cotton. The caustic soda made the cotton look more like silk, 50% stronger than before treatment and much brighter when it was dyed. Mercerized cotton, which is cotton treated with caustic soda, is still used today. Chlorine based bleaches were needed to speed up the bleaching process. The cotton industry therefore needed vast quantities of sodium hydroxide and chlorine.

HEAVY CHEMICALS

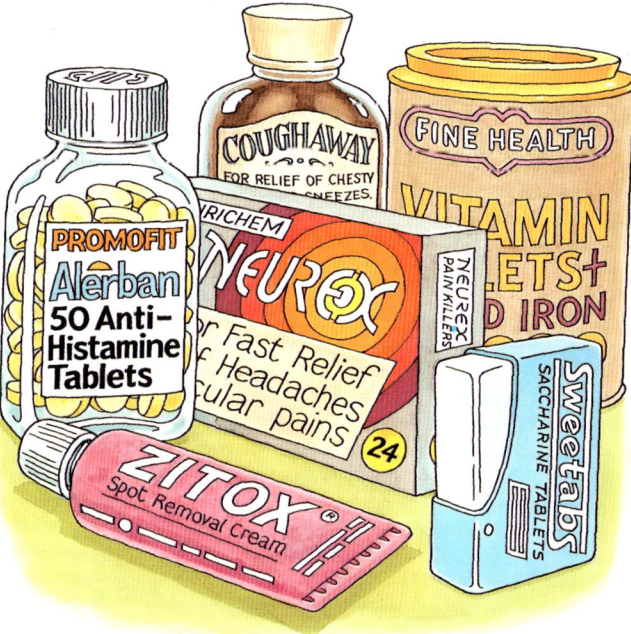

FINE CHEMICALS

The cooling and drying towers at an industrial plant for the production of chlorine

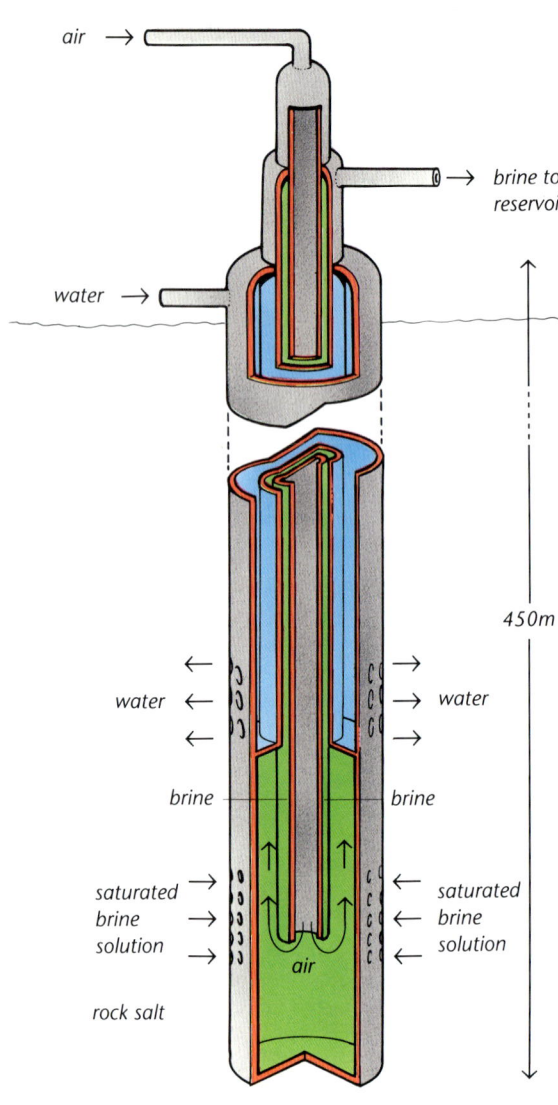

Obtaining brine from rock salt

Although Davy had carried out the electrolysis of brine about 1800, it was not until 1897 that this process was used in a factory. An American called Castner and a German called Kellner, working independently, perfected the process. In Britain the manufacture was undertaken by the Castner-Kellner Alkali Company which is now part of ICI. Their main chlor-alkali factory today is at Runcorn on Merseyside. Runcorn is near the Cheshire salt-beds and the Lancashire coal fields and has good access to road, rail and sea transport and to inland waterways.

The main steps in the production of chlorine and sodium hydroxide by electrolysis are as follows:

1 Obtaining the raw material
Water is pumped down bore-holes into salt beds. The water dissolves the salt, and the saturated brine solution is pumped to the surface. The rock in which the salt is found is insoluble and remains below the ground. In this way, little waste comes above the ground.

2 Preliminary purification
The water dissolves some impurities. These would scale up and block pipes and they are removed from the brine by adding sodium carbonate and sodium hydroxide. Calcium and strontium compounds are precipitated as carbonates by the addition of the sodium carbonate. Magnesium compounds are precipitated as the hydroxide by the addition of the sodium hydroxide.

calcium sulphate	+	sodium carbonate	$\longrightarrow$	calcium carbonate	+	sodium sulphate
$CaSO_4(aq)$	+	$Na_2CO_3(aq)$	$\longrightarrow$	$CaCO_3(s)$	+	$Na_2SO_4(aq)$
magnesium chloride	+	sodium hydroxide	$\longrightarrow$	magnesium hydroxide	+	sodium chloride
$MgCl_2(aq)$	+	$2NaOH(aq)$	$\longrightarrow$	$Mg(OH)_2(s)$	+	$2NaCl(aq)$

3 Transporting the salt

The brine is taken by pipes to the nearby chlor-alkali plant. The salt can be transported from the mine to other chlor-alkali plants by road or rail. It is expensive to transport water, so the brine is evaporated to form solid sodium chloride. It is called vacuum salt because the evaporation is carried out under reduced pressure.

4 Preparation at the chlor-alkali plant

A typical chlor-alkali plant receives brine continuously from the nearby salt mine. If the salt is brought by tanker, 140 tonnes may arrive in a day. The salt will still contain some soluble impurities. These may include heavy metal salts such iron(III) chloride. Sodium carbonate is added so that further impurities will be precipitated as insoluble carbonates. The filtrate from this is virtually ready for electrolysis.

5 The electrolysis

The electrolysis cells used in industry for making sodium hydroxide and chlorine are complicated. There are three main types, but they all produce the same three products that you produced in your electrolysis cell. All the cells are designed to keep the chlorine, hydrogen and sodium hydroxide apart from each other during the electrolysis. The diaphragm cell is the most like your electrolysis cell.

Long established chlor-alkali plants contain mercury cells, where the cathode is a shallow layer of mercury. This cell produces sodium hydroxide of high purity but has the disadvantage that mercury is expensive and toxic. Newer installations contain membrane cells. These are quite similar to the diaphragm cell but the membrane is more selective than the diaphragm.

I.   Work out

- Compare and contrast your chlor-alkali process with the industrial process. Use a whole page. Divide the page into three, vertically. Use the middle column to name the five stages involved. Use the left column to summarise each stage in your experiment and the right column to summarise each stage in the industrial process.

  Plan carefully before you begin writing!

  It might be useful to have two coloured pens, one for showing similarities and one for differences.

Purification of brine

Forty thousand cubic metres of brine are treated every day, but only half of it is electrolysed, the rest going to make salt, soda ash, etc. The solid residues are pumped back into the brine cavities for disposal.

Brine for the membrane cells has to be very pure so, as well as undergoing the above process, it is further filtered and then put through an ion-exchange process to reduce the calcium and magnesium levels to 20 parts per billion!

A battery of mercury cells for the electrolysis of brine

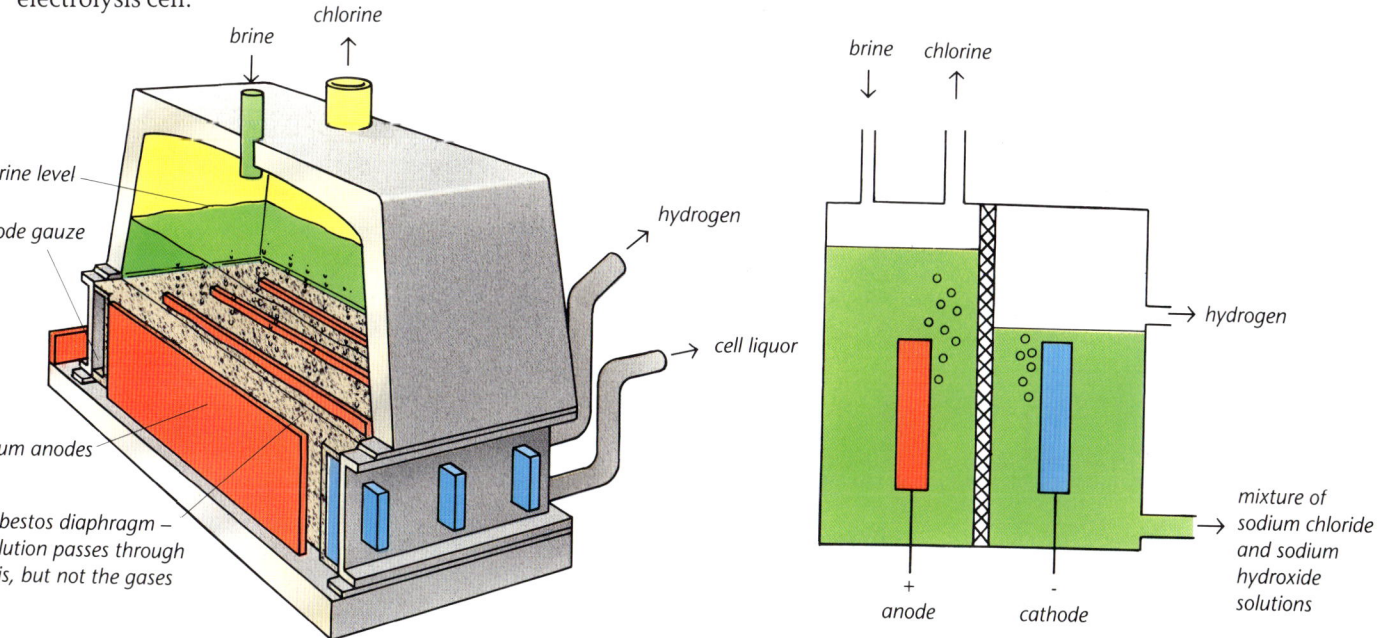

Cutaway diagram of a diaphragm cell

Simplified cross-section of a diaphragm cell

PRODUCTS FROM SALT

The diagram shows you some of the materials that are made using salt. There is quite a large number!

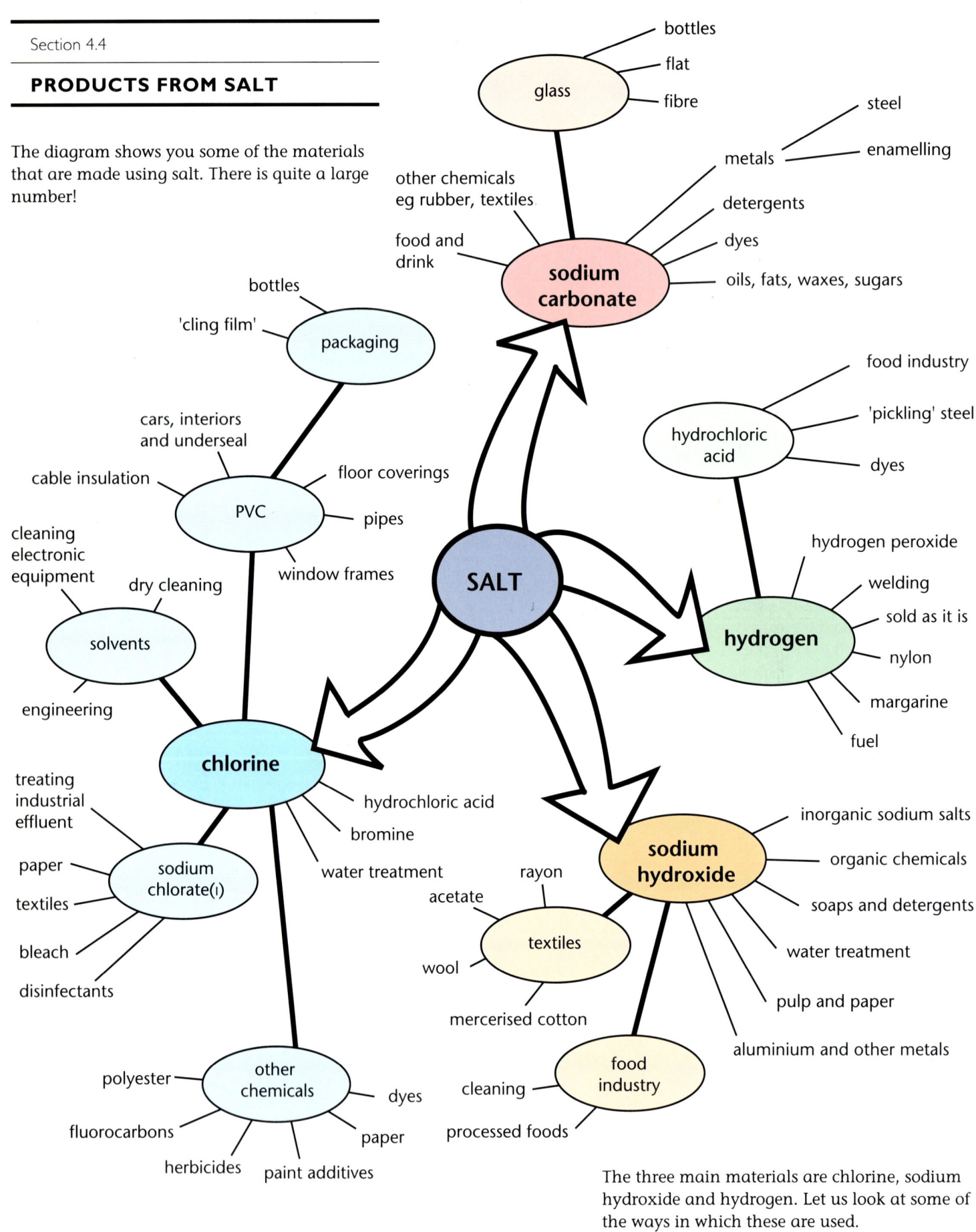

The three main materials are chlorine, sodium hydroxide and hydrogen. Let us look at some of the ways in which these are used.

Chlorine

About a third of the chlorine goes to make PVC (polyvinyl chloride). In this case, the chlorine is called an intermediate – it is used to make other chemical substances. Quite a lot of the chlorine goes to make paint. The white pigment in paint, titanium(IV) oxide, is used in a lot of paints – not just white ones. Titanium ore reacts with chlorine to form titanium(IV) chloride, which is then heated with oxygen to form the pigment. Water authorities also use quite a lot of chlorine. It is used to ensure that harmful microbes do not get into the domestic water supply.

Liquid chlorine being transported by rail

Sodium hydroxide

Sodium hydroxide is mainly supplied to factories to make other chemical substances. The cotton industry still use it for mercerising, and the wool industry use sodium hydroxide to remove oil from the wool. The paper industry is also a big customer - sodium hydroxide is used to break down the wood chips into pulp. Sodium hydroxide is used by the food industry in a number of ways, for example, to neutralise hydrochloric acid in food processing, make raising agents (sodium hydrogen carbonate is one) and also for cleaning purposes.

The interior of a mill producing paper. Sodium hydroxide is used to break down the wood chips into pulp

Hydrogen

Most hydrogen used commercially is produced by other processes. Hydrogen from this process is likely to be used as a fuel to generate steam for other chemical proceses. It may also be used to make hydrochloric acid or hydrogen peroxide. Some is used in the catalytic hydrogenation of vegetable oil to make margarine, and some is used in the manufacture of nylon.

A sample of treated oil being taken from a hydrogenation vessel. A nickel catalyst is used in these vessels to promote the reaction between the soft oil and hydrogen gas, which in turn causes the oil to harden

J. Research

- Work in a group to produce a display on the chemicals derived from salt.

  This will need planning. Some things to consider before you start are:

  Who will see it?

  Will it be on a bench or on a vertical pinboard?

  What range of items are you going to cover?

  Find out as much as you can about each item – remember the main aim is to stress the link with salt. Present the information on a card. Make it as eye-catching as possible. Try and include an amazing fact – something that is not common knowledge about your item.

  Find a sample or picture to go with your information.

UNIT 5
LIVING WITH OIL

Most ski clothes are made from synthetic materials

HOW OIL AFFECTS US

You are surrounded by many products of the oil industry – possibly more so than you think! Many clothes are now made from synthetic materials rather than natural materials such as cotton, silk and wool. A synthetic material is a material made artificially by a chemical reaction. Many synthetic materials including polyester, acrylic and nylon are made from oil.

A. Think about

- Make a list of all the articles of clothing that you are wearing.

- Cross out all the ones that contain some synthetic material.

You probably do not have many clothes made entirely from natural fibres left on your list! Oil has made an enormous difference to our lives.

In 1859, Edwin Laurentine Drake struck oil in the first well drilled specifically for that purpose. He could never have realised the tremendous impact that oil was to have on the world.

Until well into the twentieth century, chemists obtained their basic raw materials from air, water, rocks and living things. When Drake discovered oil in Pennsylvania he distilled the oil to produce paraffin. At the same time, primitive distillation of fossil fuels was also taking place in Scotland, Rumania and the Black Sea area of Russia. Paraffin from these refineries was shipped in oak barrels, each of which contained 35 gallons. If you look in the paper today you will find the cost of oil is still quoted 'per barrel'.

In the last activity you looked at some of the products that are made from crude oil. Crude oil effects peoples lives in other ways, as the following accounts show.

B. Work out

- Look carefully at each item in the display. They are all linked with oil in some way.

- Draw a table in your book similar to the one shown here.

- Write down the name of each item in the first column of the table, the substance it is made from in the second column and the reasons why we use this particular substance in the third column.

- Complete the table by writing in the last column the name of an alternative item, not produced from oil, that could be used for the same purpose.

Item	Substance	Why we use this substance	Alternative substance not made from an oil product.
jumper	acrylic	easy to care for	wool

Letter from an oil rig worker on a platform in the North Sea

Dear Becky,

When I met you last Friday, you asked me to write and tell you about my work. No, you haven't seen me around much since I left school. This is probably because for most of the time I'm living in the North Sea. (That's why I have time to write to you now!) What do I do? Well I'm working for a company based in Great Yarmouth and I'm a dive technician. For most of the time I'm away on a DSV (dive support vessel). We spend a month at a time on these. Sometimes I get to go on a shorter trip (two weeks) to one of our platforms in the southern sector of the North Sea. That's where I am now. Usually we come out here by helicopter, although sometimes we're picked up from Great Yarmouth or Lowestoft by boat.

When I'm working I'm either at the yard (that's quite tedious) or off-shore. I do 12 hour shifts (12am–12pm or 12pm–12am). Sometimes the 12 hours can be extended quite drastically, to fourteen, sixteen or more!

Between shifts I'm on call. This means that if there's a major breakdown of equipment I have to postpone my sleep. Thankfully, this doesn't happen very often.

How do I spend my free time out here? Well, it depends which rig I'm on! Some are better than others, but there's usually pool, TV, a cinema, a gym and a sauna. There is also a large turnover of people, so you have to get used to that. It's quite different from Alfield where most people live in the same house all their lives!

The pay is quite good; 'danger money' some people call it!

Well, I guess it's quite a lot compared to what you can earn in Belton's shoe shop on Saturdays. I don't get paid as much as the permanent rig personnel. But, then my advantage is that I can always move around to different rigs and vessels, so I don't get bored. As for it being danger money, I wouldn't put it like that. The safety arrangements are very thorough. All new arrivals must have induction courses and then we always have lifeboat drills once a week.

Well, I hope this has told you something about what I do. See you in a fortnight,

Love Mike

A pensioners' account of life in the Shetland Islands

I count myself fortunate to have lived in the Shetland Islands for 70 years. My family have lived here for generations, too many to count. There are about a hundred islands and they are on the same latitude as Oslo and Leningrad. What I most love about the islands is the sea – I don't think I can travel anywhere on any of the islands and be more than five miles from the sound of the waves lashing the coast. We are a close knit community, with most of the adults employed in just four industries – agriculture, knitwear, fishing and fish processing. A few smaller crafts like brooch making and pottery employ a few. On the whole, we are not a rich people – our richness comes from our way of life. But there is almost full employment and wages are adequate.

Well, that was what it was like before 1972 – then our lives changed. I can only say for me that it was for the worse! Seismic explorations had indicated that quite sizable oil reserves were situated just off our islands. I, like many of the islanders, thought that, at the worst, the islands would serve as a service base. We would get new quays and harbours, a few new jobs and after all the fuss was over the fishermen would be grateful to have better landing conditions. I seriously underestimated what was to happen. In 1972, Shell announced the discovery of the giant Brent field. Doom and gloom soon set in. Was land to be the subject of compulsory purchase? Would our way of life change forever? Would our beautiful landscape and the associated ecology be destroyed? Would there be massive oil slicks so that fishermen could not fish? These thoughts went through my mind and I am sure I was not alone in having these thoughts. We should not have been surprised by the influx of visitors. Many were speculators trying to buy land. Others just wanted to see Shetland as they knew it – like us they were sure it would never be the same again.

It seemed that things changed almost overnight. The population increased by 10 000 in just 10 years. (The population in 1971 was 17 500). New houses were needed and new schools. More roads had to be built and all the other services improved to cope with a near doubling of the population. The oil industry paid higher wages than the traditional industries of our islands. It was not unexpected when the fish processors and knitwear producers had trouble finding sufficient labour – they couldn't

Making woollen garments is a traditional craft in the Shetland Islands

pay the same wages so everyone tried to get jobs in oil. People in 'the oil' were now earning in a month what they had earned previously in a year. I think the problem really hit me when I read that a special study was to be undertaken to investigate ways of ensuring that the bakeries still functioned with such a shortage of labour. The Shetland Times commented that 'the problems of affluence might be greater than poverty'; there is much truth in that. Previously we didn't really have a class system on these islands. We do now. There are those that have money and those that do not and the gap between us is large.

There are other problems. No-one that I knew ever used to lock their car or their house. We do now. There is theft and vandalism and an increase in alcoholism. This worries me – a lot of youngsters are drinking. They have money in their pockets and seem content to spend it on drink.

What will happen when the oil runs out? Well, I don't think things will ever be the same again. I hope that more thought goes into providing for the islands future than went into planning for this so called oil bonanza. Some things have improved – there are now more social activities, clubs and the like for drama and so on. Youngsters can now find work on the islands and don't feel so inclined to leave for the mainland. But, then neither of these bonuses are likely to stay when the oil has gone. Will it seem like a ghost place? I hope not. I have had a good life, the islands are all I have known and I would like to think that future generations will benefit from this so called bonanza. I feel we have paid dearly for it.

VIOLET HARRIS

A journalist's report on a bad oil spill

A massive oil slick now covers an area of at least 240 square miles. To date it is the world's worst oil pollution disaster. It is impossible to visualise, but take out a map of your area and draw a box 12 miles by 20 miles and then you will appreciate the size of the problem that has to be tackled. The slick is not static. At the moment it is moving at about a mile an hour. Many countries in this area need water from this area of sea. They are dependent on the desalination plants providing them with fresh water. If the slick reaches the desalination plants' intake, they will be blocked and the plants will become inoperable. Many experts believe that technology cannot clean up a slick this size.

The slick is seriously affecting the marine life and the local fishing industry. Already I have seen thousands of dead seabirds, fish, turtles and dugongs (sea cows). More animals will die – they are dependant on the plankton. Now also is the breeding time for shrimps – their larvae can not cope with the oil. If chemical oil dispersants are used these can be even more harmful than the oil to marine life. Research in the UK has shown that these prevent shrimps finding their food and prevent males finding females. The fishermen expect the valuable oysters to suffer too. They are despondent, because this slick is likely to ruin their livelihood.

What is to be done? One way of trying to deal with the oil is to suck it up! This involves directing the oil with giant booms or skimmers. A recovery ship then removes the oil laden water. The oil is skimmed off and pumped to an adjacent storage vessel, and the treated water is returned to the sea. Alternatively, absorbent booms of nylon or wood chip pads are held by nets. These can absorb up to twenty five times their own weight in oil. Biotechnology may also play a part. Bacteria have been found in aircraft fuel tanks, digesting the fuel. This fact is now being put to good use. These bacteria eat the oil and break it down. If preparations of bacteria are sprayed on the oil slick they break it down and excrete minerals to the sea. These can be used by sea grasses, favourites of the endangered dugongs.

Coordination of this huge operation is essential. Many experts are giving advice on this tremendous disaster. We all hope that those that claim 'the ecology of the area will be destroyed for decades' will be proved wrong.

JABER AL-SALEM, MIDDLE EAST

A Saudi journalist looking at a pool of oil on a polluted Persian Gulf beach near the Kuwait – Saudi border, January 29 1991

You have just read how three people's lives are affected by oil. Your life is also affected by oil, but perhaps not in such a dramatic a way.

C. Think about

- Describe a day in your life and how it is affected by oil. Think carefully about what you are going to write before you begin. If you can, get someone over 60 to tell you what they used instead of oil products when they were your age. This will help you to understand how oil affects your life. Write your account as an article for a school magazine, local paper or as an entry for a competition for writing about science issues. The best articles will be those that are well researched and planned.

USING OIL

James Young (1811 – 1883)

When crude oil comes out of the ground, it is of limited use. It is a complex mixture of compounds. Many of these compounds only contain the elements carbon and hydrogen and are called hydrocarbons. Because the oil was formed from living organisms, it also contains small amounts of some of the other elements which are present in living things, such as oxygen, nitrogen, sulphur and some metals.

Oil has been used for thousands of years, but for most of this time it has been used just as it was found. In the middle of the nineteenth century, James Young was asked by the owners of a Derbyshire coal mine to investigate the oil that was seeping into the mine shafts. He was interested in this liquid and tried heating some of the oil, in similar apparatus to that being used for coal distillation at the mine. He obtained paraffin. Later, paraffin was burnt in paraffin lamps and provided a much needed improvement on previous lighting. Oil then

began to be distilled on a larger scale to provide the necessary paraffin for lighting and heating. Some lubricating oil was also separated out. The rest of the crude oil mixture was regarded as waste, and either burnt or allowed to drain into the ground.

When the internal combustion engine was invented in 1885 petrol was needed. However, it was 1912 before petrol became a more important crude oil product than paraffin. During a visit to America in 1872, the Russian chemist Mendeleyev commented that oil was too valuable to be burnt. He suggested that it was a potential source of organic chemicals. Organic chemicals are those containing carbon, and they are very important in our lives. Plastics, soaps, drugs and dyes are all organic chemicals. The development of these chemicals from oil really began in America during the second world war. The industry that developed is now called the petrochemicals industry and is one of the world's major industries. It is estimated that over seven million different organic compounds have been made.

Separating the crude oil mixture into a number of different parts is still the basic process that goes on at an oil refinery. When crude oil is heated it does not have a fixed boiling point because it is a mixture of many components. Each component boils and turns into a gas at a different temperature. This fact is used to separate them by the process of fractional distillation. The different components are called fractions. In general, the first fractions to boil have the smallest molecules. Those with between one and four carbon atoms in their molecules are gases and boil below 0°C.

Fractional distillation is carried out in steel column which may be 50 metres tall. The diagram shows you what happens during the process. An oil refinery may separate up to 10 million tonnes of crude oil each year in a continuous process.

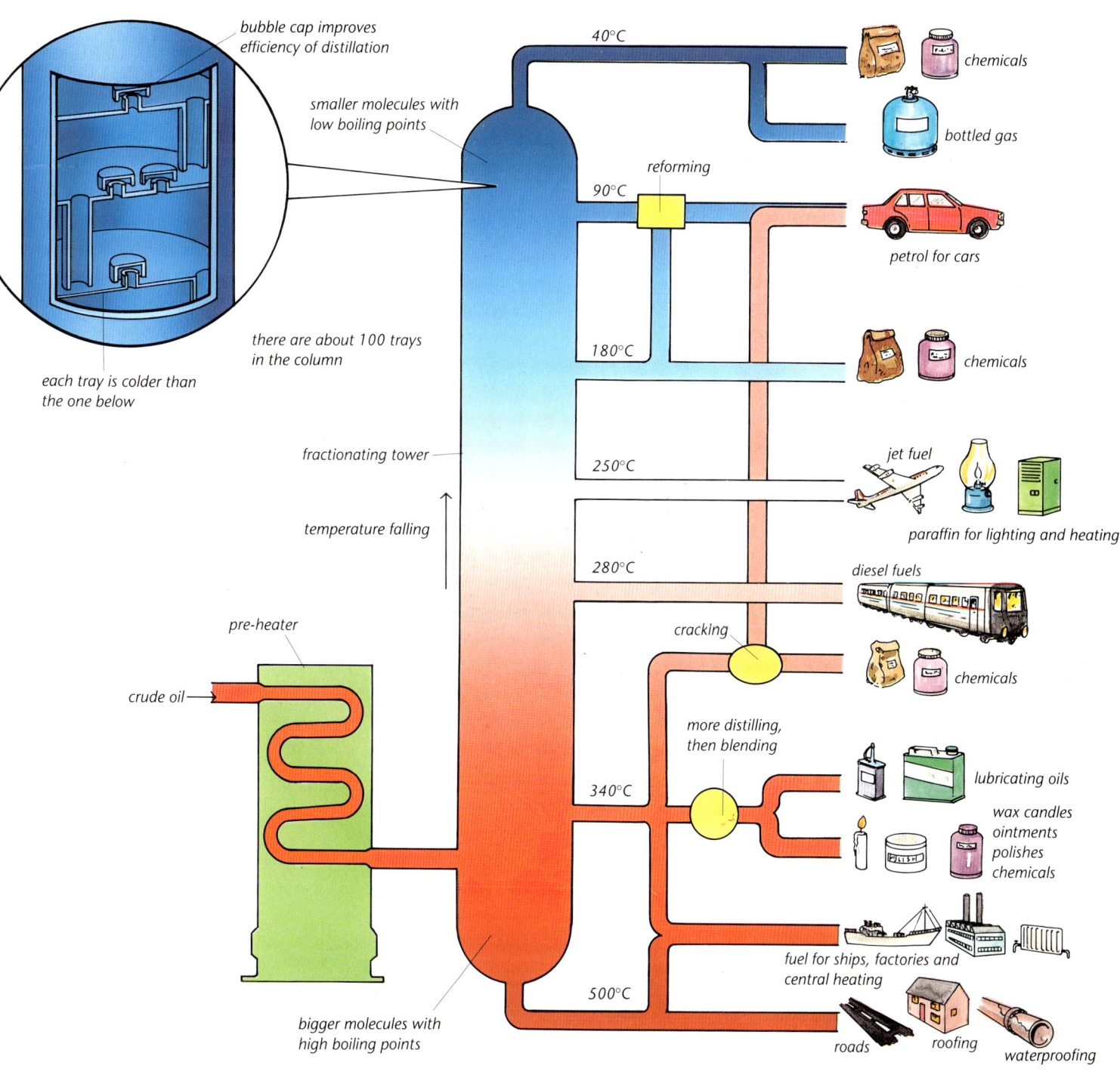

bubble cap improves efficiency of distillation

smaller molecules with low boiling points

each tray is colder than the one below

there are about 100 trays in the column

fractionating tower

temperature falling

pre-heater

crude oil

bigger molecules with high boiling points

40°C

90°C   reforming

180°C

250°C

280°C

340°C

500°C

cracking

more distilling, then blending

chemicals

bottled gas

petrol for cars

chemicals

jet fuel

paraffin for lighting and heating

diesel fuels

chemicals

lubricating oils

wax candles
ointments
polishes
chemicals

fuel for ships, factories and central heating

roads   roofing   waterproofing

Not all crude oil is the same. The following table shows the percentage (by mass) of different fractions in crude oil from different parts of the world.

Source	Gasoline	Naptha	Kerosene and gas oil	Fuel oil
UK North Sea	15	15	40	30
Nigeria	10	15	37	38
Middle East	8	12	32	48
North Africa	15	15	30	40
Venezuela	1	1	19	79

Composition of different crude oils (% by mass)

As a rough guide, crude oil with lighter fractions, such as naphtha, is more valuable than crude oil with the heavier fractions, such as fuel oil.

D. Think about

- Which countries appear to own the most valuable crude oil? Explain your answer.

E. Research

- The 'demand barrel' shows the relative demand for different oil products in Europe in 1990.

  What do you notice about the demand in Europe compared to the actual content of the various types of crude oil given in the table?

- Look in a newspaper (in the financial section) to find the cost of a barrel of crude oil. The price will be given in $. By looking at the exchange rate you can convert this to £. Monitor the cost of a barrel of oil over the next few weeks.

  What factors might be affecting the cost?

others
naptha
kerosene
LPG
fuel oil
gas oil
gasoline

Demand barrel for Europe

The demand for the different fractions from crude oil does not match exactly any one type of crude oil. Also, the demand for different fractions varies. In summer more gasoline/petrol is needed but in winter more fuel oil is required. It is in a refinery's interest to meet these demands.

Fractional distillation separates the different components in oil. It does not involve any chemical reactions. In the next stage of the refining process the components are changed to produce new substances. It is these chemical processes of cracking and reforming which enable the refinery to produce fractions to suit the demand.

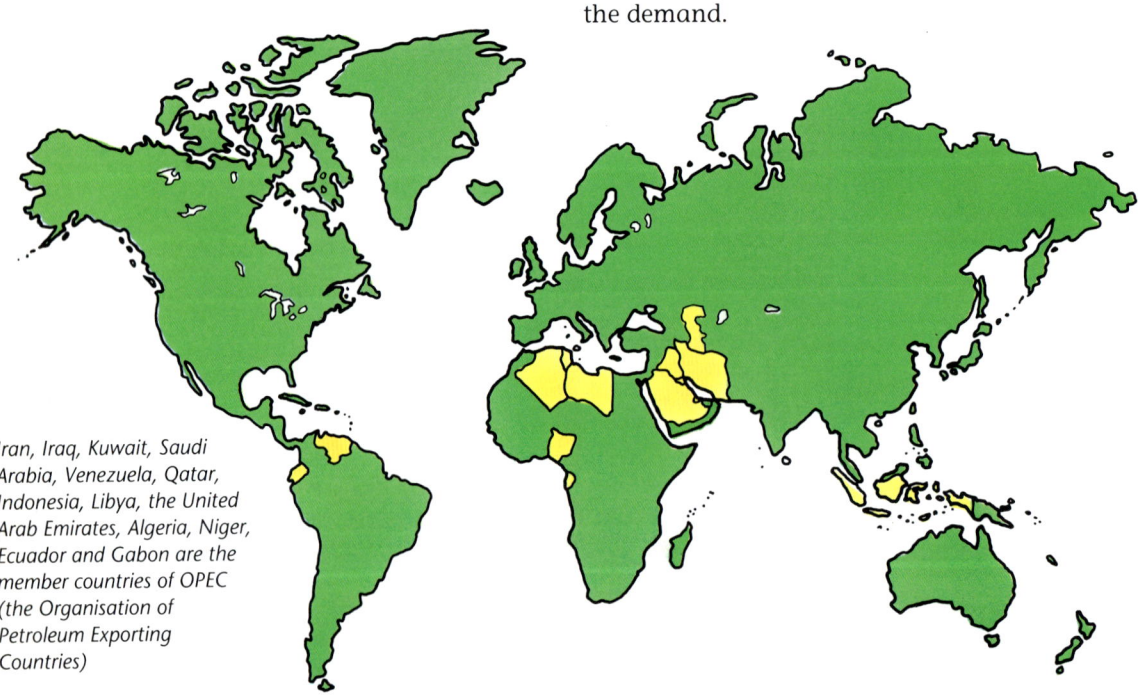

Iran, Iraq, Kuwait, Saudi Arabia, Venezuela, Qatar, Indonesia, Libya, the United Arab Emirates, Algeria, Niger, Ecuador and Gabon are the member countries of OPEC (the Organisation of Petroleum Exporting Countries)

Cracking

In cracking, large molecules are broken or 'cracked' into smaller molecules, with fewer carbon atoms. The fractions containing smaller molecules boil at lower temperatures and tend to be more valuable than those containing bigger molecules. The large molecules form the feedstock for the process. At present naphtha and gas oil are frequently used, but in theory any hydrocarbon can be used. The cracking will take place if the feedstock is heated with steam to a temperature of 800°C. The heat is needed to break the carbon-carbon bonds in the naptha. If a catalyst is used, lower temperatures of about 500°C can be used. The feedstock is vapourised and enters the cracker along with a powdered silica/alumina catalyst. After the catalyst has been cleaned it can be re-used. The products obtained depend on the operating conditions and the feedstock.

Reforming

A number of different compounds, each made up of small molecules, are formed during cracking. These are separated in an adjacent gas separation plant. Many of these compounds undergo a further reaction in a reforming stage. In this reaction, the structure of the molecules is changed, or small molecules are combined to make the precise type of larger molecules in demand. One important product of Shell's large Stanlow cracker complex in Cheshire is a heavier hydrocarbon formed from two different types of liquid petroleum gas (LPG). This is used in the production of unleaded gasoline. By the complementary processes of cracking and reforming, products can be produced to match the demand.

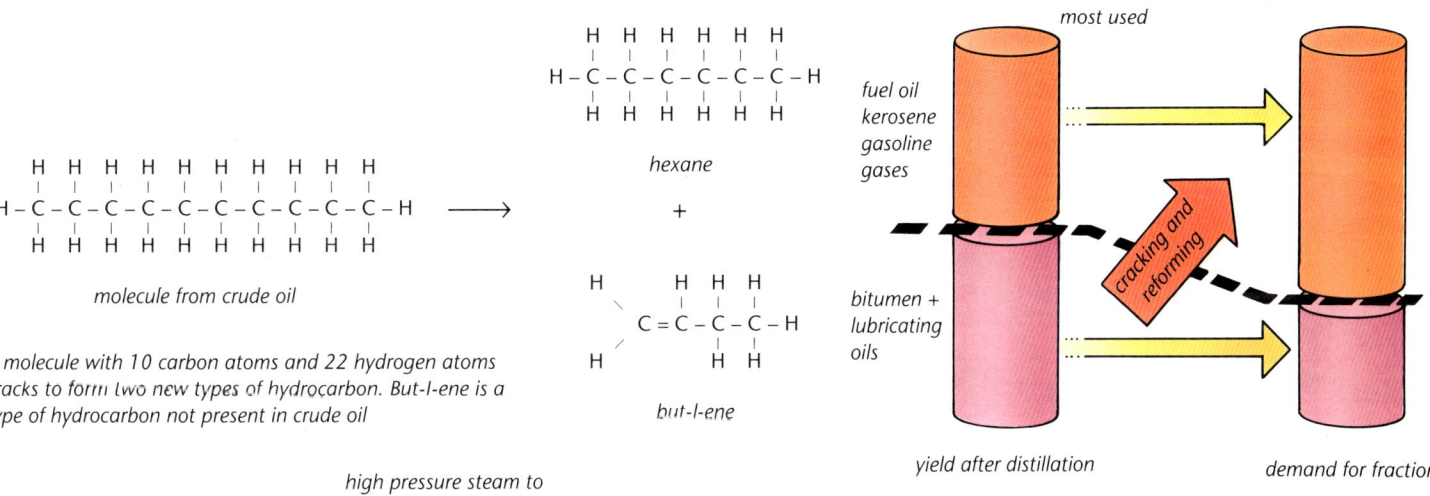

A molecule with 10 carbon atoms and 22 hydrogen atoms cracks to form two new types of hydrocarbon. But-I-ene is a type of hydrocarbon not present in crude oil

molecule from crude oil

hexane

+

but-I-ene

Cracking and reforming produce the smaller fractions which are most used

fuel oil
kerosene
gasoline
gases

bitumen +
lubricating
oils

cracking and reforming

yield after distillation

demand for fractions

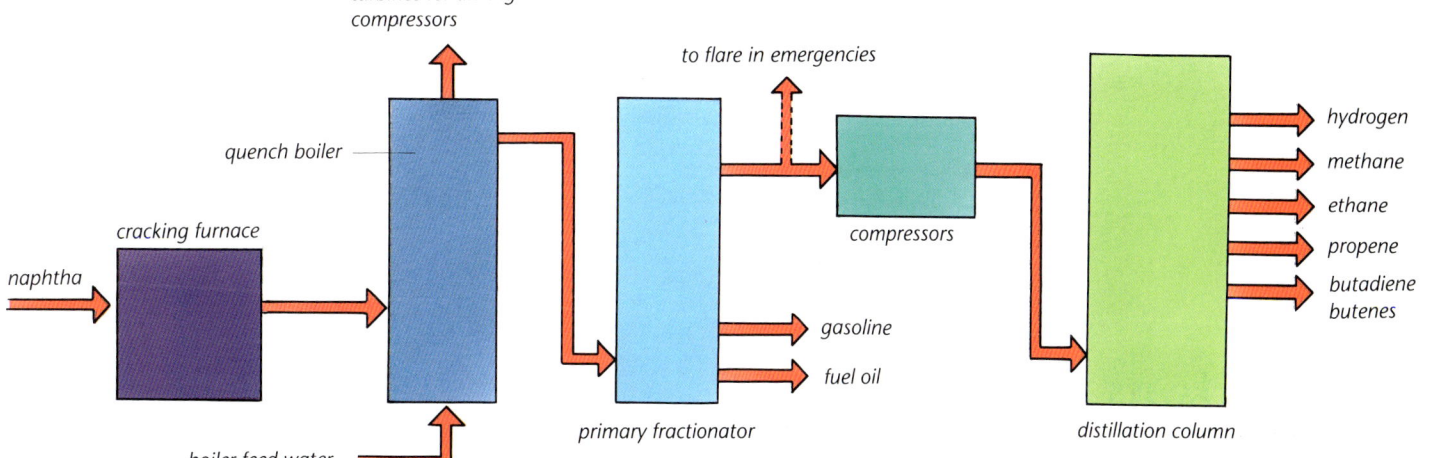

high pressure steam to turbines for driving compressors

to flare in emergencies

naphtha

cracking furnace

quench boiler

boiler feed water

primary fractionator

gasoline

fuel oil

compressors

distillation column

hydrogen

methane

ethane

propene

butadiene
butenes

Aerial photo of an oil refinery in Louisiana, USA

F. Observe and record

- Look carefully at the photograph:

  1 Explain how the oil arrives at the refinery.

  2 Identify:

     a the storage tanks for the crude oil;

     b the fractional distillation unit;

     c the catalytic cracker plant;

     d the tall vertical tanks for storing chemicals;

     e the special spherical storage tanks for LPG.

  What other features do you recognise?

G. Work out

- Draw a flow chart to show the path of a carbon atom, as it journeys from being in the crude oil deep under the North Sea, until it ends up first in unleaded petrol and then in the atmosphere. Include as many processes and technical terms as possible. You could also include diagrams.

THREADS, WEBS AND POLYMERS

A village woman in Thailand spinning silk from cocoons

A spider's web

As long ago as 1665 Robert Hooke suggested that it should be possible to copy the threads that animals such as silk worms and spiders make, and produce synthetic fibres. However, because of production difficulties, synthetic fibres have only been made on an industrial scale since 1919. These synthetic threads are polymers. Polymers are long chain molecules made up of thousands of small molecules. These small molecules are called monomers and, in many cases, all the momomers in a polymer are identical. If there are several different type of monomers the large molecule is called a co-polymer.

a polymer model

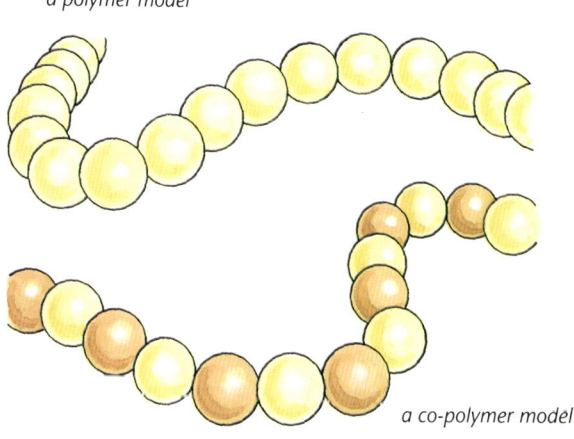

a co-polymer model

These are models of two polymers! The word polymer comes from two Greek words, poly *meaning 'many' and* meros *meaning 'parts'*

Synthetic polymers are often wrongly called plastics. Many polymers are fibres which are not plastics. The word plastic means capable of being easily moulded. Plastics are polymers, but not all polymers are plastic.

The first synthetic polymers were made by modifying natural polymers. It is possible to make a polymer from casein, the protein in milk. Your grandparents may have some items made from this polymer at home. It is a creamy white material.

Rolls of silk in a shop in Thailand

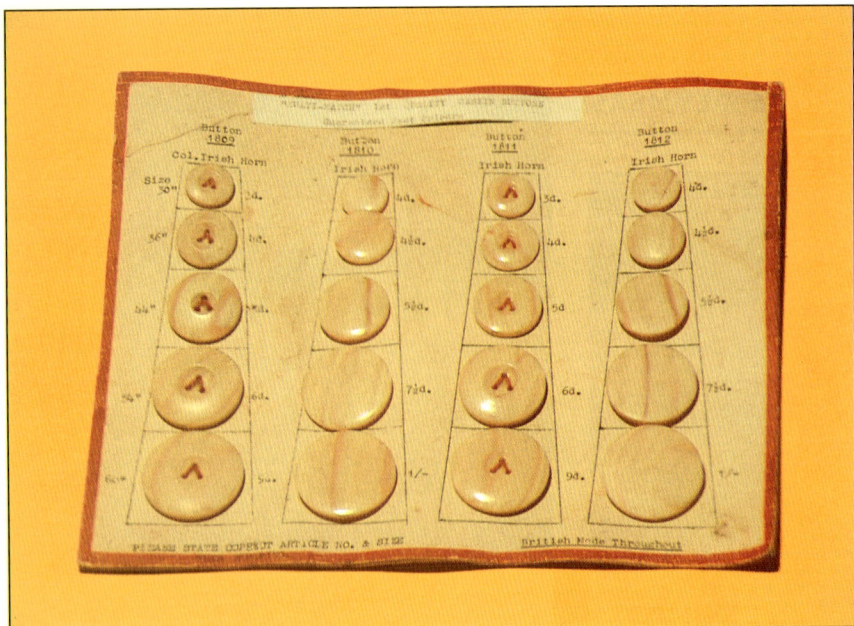

Earlier this century casein was used for objects such as knitting needles, buttons, crotchet hooks and knife handles

Today, synthetic polymers are made from small molecules obtained from crude oil. These monomers can be put together in combinations that do not occur naturally. It is possible to design polymers that have suitable properties for a whole range of jobs.

| | Feedstock | | | | |
Product	Ethane	Propane	Butane	Naphtha	Gas oil
Hydrogen	10	3	2	1	1
Methane	6	28	24	15	8
Ethene	76	43	39	30	23
Ethane	ethane is recycled and cracked				
Propene	3	16	15	16	14
Propane	ethane and propane are recycled and cracked		1	1	1
Butenes	1	2	7	5	5
Butadiene	2	3	3	5	6
Gasoline	2	4	7	23	20
Fuel oil	–	1	2	4	22

Products obtained from cracking various feedstocks (expressed as percentage by mass)

H. Work out

- The table shows the products obtained from cracking different feedstocks at an oil refinery. What can you conclude about the products when different feedstocks are cracked?

One important compound formed by cracking is ethene.

$$H \quad \quad H$$
$$\backslash \quad \quad /$$
$$C = C$$
$$/ \quad \quad \backslash$$
$$H \quad \quad H$$

the formula for ethene

The polymer made from ethene monomers is polyethene. It is more commonly called polythene. Polythene is made when ethene molecules join up. In a polythene bag each polymer of polythene would be made up of between 30 000 to 40 000 monomers!

$$\begin{array}{ccccc} H & H & H & H & H \\ | & | & | & | & | \\ -C & -C & -C & -C & -C- \\ | & | & | & | & | \\ H & H & H & H & H \end{array}$$

the formula for polyethene

I. Make

- Use a molecular model kit to make an ethene molecule.

- Make two more molecules of ethene, or join up with two other groups to make a model of a part of a polythene molecule.

- Describe how you made a part of a polythene polymer. You should include the words 'bonds', 'monomer' and 'polymer' in your description. It will probably help to draw some diagrams. Make sure you indicate the repeating monomer units.

Polythene is an example of an addition polymer. It is the double bond in the small ethene molecule that is important. Polythene is formed by the process called addition polymerisation when the monomers just add onto each other.

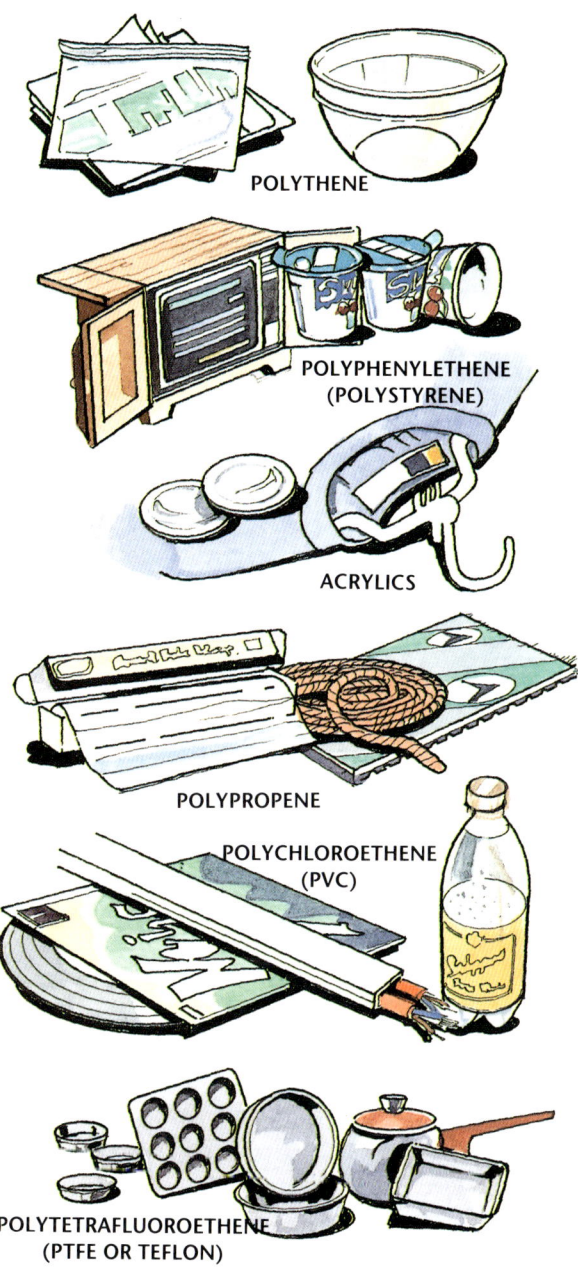

POLYTHENE

POLYPHENYLETHENE
(POLYSTYRENE)

ACRYLICS

POLYPROPENE

POLYCHLOROETHENE
(PVC)

POLYTETRAFLUOROETHENE
(PTFE OR TEFLON)

Some examples of addition polymers. Collect examples of addition polymers to make a display

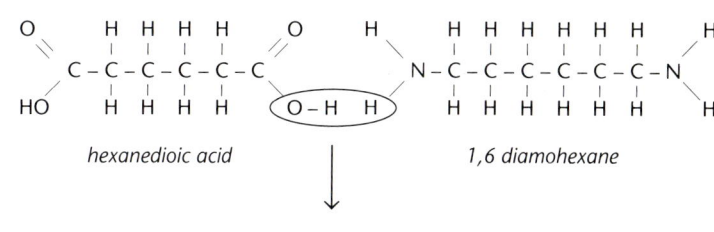

hexanedioic acid   1,6 diamohexane

2 units of a nylon polymer

... or more simply

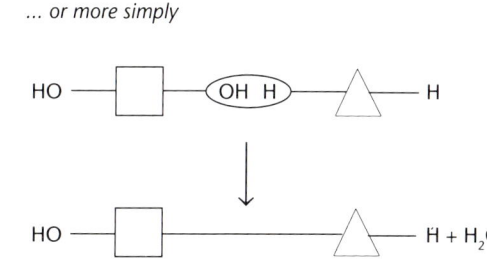

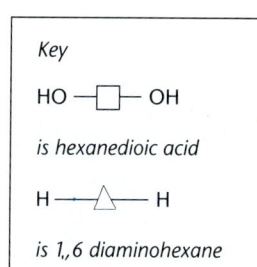

Key

HO —☐— OH

is hexanedioic acid

H —△— H

is 1,,6 diaminohexane

The family of polyamides which include nylon are examples of condensation polymers

---

J. Work out

- Make up your own symbols and draw a simple diagram to show what happens when the condensation polymerisation reaction shown below takes place.

- Make a chart which shows the similarities and differences between addition and condensation polymerisation.

Not all polymers are made just by joining up the separate monomers. Some polymers are made by condensation polymerisation. In condensation polymerisation there are usually two different sorts of molecules that provide the two different groups that are needed for the reaction. During polymerisation these two different groups react and a small molecule is given out. The small molecule is often water – which may help you to remember that it is called condensation polymerisation. (The small molecule could, however, be another substance, for example hydrogen chloride, HCl).

1,6 diamohexane   hexanedioyl chloride

+ HCl

2 units of a nylon polymer

NYLON FAMILY

POLYURETHANE FAMILY

POLYESTER FAMILY

EPOXY FAMILY

Some examples of condensation polymers. Collect some examples of condensation polymers for a display

'LIQUID GOLD'

Fossil fuels, such as crude oil, take millions of years to form and so cannot be replaced. In 1989 the projected world reserves of crude oil were 907 billion barrels, of which 74.5% were in OPEC countries. The following pie charts show the world reserves of all fossil fuels and what the crude oil is used for in Europe. At present rates of use, oil will become increasingly scarce. That which remains will be difficult to extract and so very expensive to use. Alternative raw materials, fuels and other products will need to be considered.

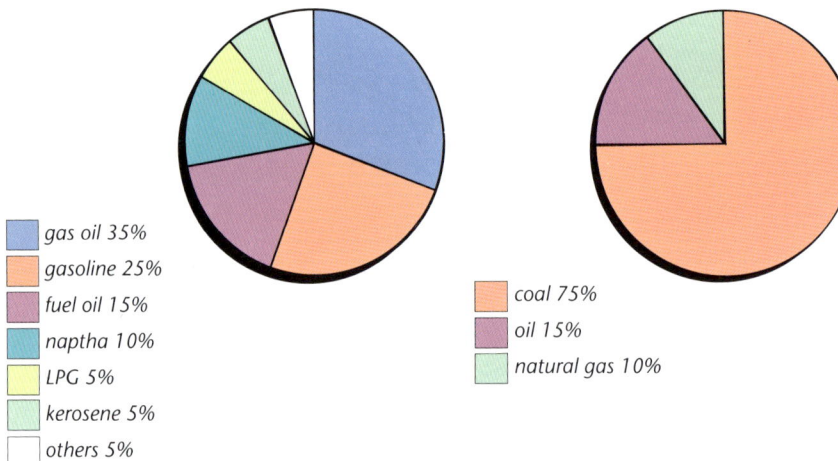

- gas oil 35%
- gasoline 25%
- fuel oil 15%
- naptha 10%
- LPG 5%
- kerosene 5%
- others 5%

- coal 75%
- oil 15%
- natural gas 10%

What the crude oil is used for in Europe   *World reserves of fossil fuels*

K. Think about

- Design a survey to find out how much your friends and family know about plastic polymers. Plan it carefully – people may find it much easier to answer a multi-choice question than to think up an answer from scratch. You may be able to show the people samples of polymers and ask questions about these.

  Bear in mind how you will analyse and present your results.

- Try out your questions on a friend in class. When you are satisfied with the questions try them out on at least ten people (Not those in the class with you.)

- Make a presentation of your findings.

Here is one view from the petrochemicals industry about how to make the best use of the remaining oil:

'In the future, using alternative fuels will allow oil to be conserved for manufacturing chemicals, including plastics. For example, it will be more cost-effective if other raw materials, such as coal, are mined in greater amounts as a fuel supply. With these facts we can be sure that the plastics industry will not have a real problem for hundreds of years.'

This is a very optimistic view and if all branches of the oil industry had the same view these would definitely be a problem!

What will happen when the oil becomes too expensive to use? You may like to look back to Section 1 of the unit to remind yourself of how you depend on oil.

The following is a result of a quick brainstorm that a class did. You may find it helpful as a starting point for the work out activity, but do alter it and add to it to suit yourself.

In OECD Europe, gasoil commands a 35% share of the demand barrel, since it is widely used for domestic heating, especially in Germany, and since diesel cars are comparatively popular in Europe

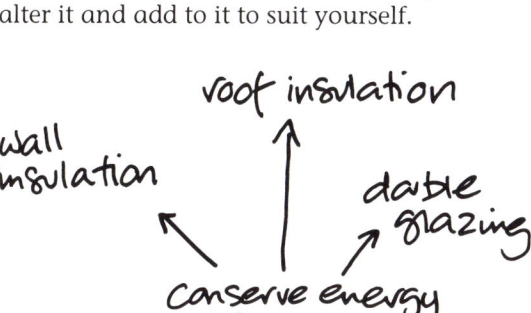

wall insulation

roof insulation

double glazing

Conserve energy

transport

use more public

Singapore roads

better designed cars

use alternatives to oil now → nuclear fuel
→ coal
→ biomass
→ others → solar
→ wave
→ wind
→ geothermal
→ hydroelectric

save/recycle plastic polymers → wash & re-use
→ change use eg. plastic grass
↓ melt & remould
burn & use heat

L.   Work out

- You are to contribute to a special supplement for a newspaper called 'When oil becomes too expensive to use – what can we do now?'.

- Consider who your audience is (you may be able to choose your ideal audience). The articles must contain data and still be interesting and relevant. Your audience must know how they can help and not just believe it is a problem for other people. Try and think up some ways of getting your audience to do something about the problem – rather than just read and then forget about it.

- You will need to plan the supplement carefully so that the articles are balanced. It is important that you sort out the tasks before you begin so that you avoid duplication (or omissions).

- Worksheet CPr23 When oil is too expensive to use – some research contains some information that will help, but you will need to do as much research as possible.

UNIT 6
THE FERTILISER INDUSTRY

It seems that hardly a day goes by without a mention in the news of a famine in one part of the world or another. The facts behind the headlines are staggering. It is estimated that, in the world as a whole, one person in eight does not have enough to eat and possibly as many as two thirds of the people cannot get the right types of food. This is the situation at the moment. The world population is increasing rapidly, especially in developing countries. Unless something is done, the situation will get worse. Humans, like all animals, depend on plants for their food. It is plants that can turn solar energy into carbohydrates and other essential substances, using minerals from the soil. If we are to produce more food, our attention must go to improving the crops that are grown. One way of improving crops is to increase their yield by using fertilisers.

But what exactly are fertilisers? In the early 1800s a number of French chemists were studying plant growth. De Sassure found that although plants could make their own carbohydrates using carbon entirely from the air, they also needed to take in a number of substances from the soil in order to grow well. Liebnig considered that the fertility of the soil would only be maintained if these substances were added back to the soil. By careful analysis of the ash of plants, he devised his own fertiliser containing mostly phosphorus and phosphate salts. Unfortunately it did not work at all well because he did not include any salts containing nitrogen. He assumed that plants got this directly from the air.

Gilbert and Lawes studied plant growth and by 1855 had realised the importance of fertilisers containing nitrogen. Warrington, working with Lawes, discovered the importance of microbes in the soil in turning ammonium compounds into nitrates that could be taken in by plants. Vital links in the nitrogen cycle were added. Farmers had used urine and dung for many centuries to improve plant growth. It was now known how nitrogen in these compounds was absorbed by plants.

The fertiliser industry was beginning to take off. Lawes set up a fertiliser factory in Deptford, South London. A variety of fertilisers were being developed, including 'superphoshate'. This was made by adding sulphuric acid to phosphate originally derived from bones, but later obtained from rock. Ammonium sulphate, a waste from the coal gas industry, was used extensively after 1850. Guano (dehydrated bird droppings) was imported from Peru in the early nineteenth century. More and more fertiliser was needed, as its usefulness became clear. Sodium nitrate was discovered in Chile and this too was shipped to Britain.

Although the deposits of sodium nitrate in Chile were large, many argued that a new source of nitrogen had to be found. Nitrogen was known to make up 80% of the air.

Graph showing how the yield of winter wheat is affected by using nitrogen fertiliser

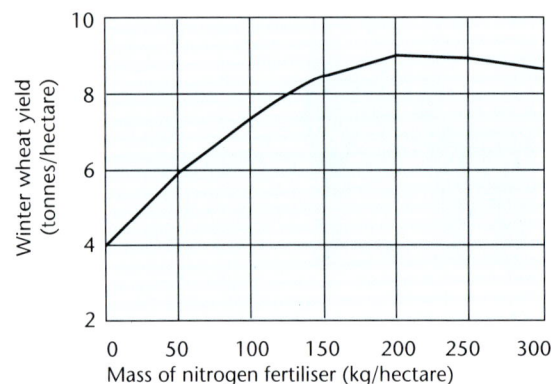

Fertiliser being loaded ready for spraying

Chemists worked hard to find a way of turning this nitrogen into compounds that could be used. Unfortunately, nitrogen is unreactive.

Around the beginning of the twentieth century, two processes were developed which could use the nitrogen from the air to form a compound (in a process called fixing). Unfortunately, both had drawbacks, one of the main ones being that they needed a large supply of cheap electricity and this was not available at that time. It was Fritz Haber who, with his painstaking theoretical work, solved the nitrogen-fixing problem. He studied the reaction between nitrogen and hydrogen to form ammonia.

The reaction can be represented as:

$$N_2(g) \ + \ 3H_2(g) \rightleftharpoons 2NH_3(g)$$

nitrogen   hydrogen   ammonia

Energy is given out during the reaction. This energy causes the surroundings to get warmer. The reaction is said to be exothermic.

In practice this reaction reaches an equilibrium. This means that the nitrogen and hydrogen are reacting to form ammonia at the same rate as the ammonia is decomposing to form nitrogen and hydrogen. Changing the conditions can increase the rate of either of these two reactions.

Haber knew that:

1   At ordinary pressures hardly any nitrogen and hydrogen would combine, but increasing the pressure would increase the yield of ammonia.

2   The reaction was exothermic and so a low temperature would favour a higher yield of ammonia. However, this would increase the time needed to produce the ammonia. A compromise was therefore needed.

In 1908, after working on this problem for four years, Haber found that he could get an 8% yield of ammonia at a temperature of 500–600°C and a pressure of 175 atmospheres if he used a suitable catalyst (he used osmium).

The next problem to tackle was how to turn this laboratory process into a viable industrial one.

Haber suggested two important features:

1   Continuous flow, in which the reactants are continually supplied to the process and the product continually removed.

2   Heat exchange, in which the heat from the ammonia leaving the reaction is used to warm up the reactant gases.

Not only are these two ideas still used in the production of ammonia but they are also used in many other industrial processes.

It was Carl Bosch, who at that time was working for a dyestuff manufacturer in Germany, who solved the complex problems of producing the apparatus on an industrial scale. By 1912 such a plant was producing sufficient ammonia to allow the production of 30 tonnes of ammonium sulphate a day. Nobel prizes

Carl Bosch (1874 – 1940)

Today ammonium based fertilisers (and nearly all nitrogen compounds) are made from ammonia which has been synthesised by the Haber process.

Fritz Haber (1868 – 1934)

A. Work out

- The table on this page shows a number of different fertilisers that are used. Use this information to help you to complete the table on Worksheet CPr24.

Fertiliser	Formula
*ammonium nitrate	$NH_4 NO_3$
*diammonium hydrogenphosphate(v)	$(NH_4)_2 H PO_4$
*ammonium sulphate	$(NH_4)_2 SO_4$
tricalcium phosphate(v)	$Ca_3 (PO_4)_2$
potassium chloride	KCl
potassium nitrate	KNO_3
tripotassium phosphate(v)	$K_3 PO_4 . H_2O$
sodium nitrate	$NaNO_3$

A fertiliser producing plant

B. Make

- Working in a group, make a sample of a fertiliser. First, decide which fertiliser you will make. Worksheets CPr25, 26 and 27 provide instructions for making those fertilisers marked in the table with an asterisk. Discuss your plans with your teacher, before you begin.

D. Investigate

- Plan, carry out and evaluate a study of the effects of fertilisers on plant growth. The worksheet *Planning an investigation* will help you.

C. Research

- Compare the process by which you made your fertiliser with that used in industry.

- Find out about the economic, environmental and health and safety factors involved in the industrial manufacture of this fertiliser.

- Write a report on the manufacture of the fertiliser. There is a checklist on Worksheet CPr28 to help you consider all possible factors.

PATTERNS OF BEHAVIOUR

Contents

A WORLD ON THE MOVE

Section 1.1

THE KINETIC FUNFAIR

All the chemical substances in the universe are called 'matter'.

These substances can be grouped as solids, liquids or gases. These groups are called the three states of matter.

How do they make candyfloss so big with so little sugar?

Why can't I suck the ice cube up the straw?

Why does the drink go everywhere when I spill it?

A. Discuss

- As a group, identify as many solids, liquids and gases in the picture as you can.

Jamie took his six year old brother, Joe, to the fair. He bought him a drink of orange and a candy floss but rather wished he had not – Joe was full of questions. Three of his many questions were about solids, liquids and gases.

B. Plan

- Plan three simple experiments that Jamie and Joe could do at home that would help Joe to understand more about solids, liquids and gases. Explain what each investigation is trying to show. You could do this in the form of a conversation between Jamie and Joe.

- If possible try out your experiments and explanations on young children, to see how clear your explanations are.

In the period around 300 – 400 BC, the Greeks attempted to explain the whole world and to show how everything in it had formed. One of the important ideas from this period was that there were four basic elements: earth (solid), water (liquid), air (gas) and fire (a source of heat and energy). The first three of these elements are very similar to the three states of matter that we use today for classifying substances. The Greeks of this period also tried to describe matter. Democritus came very close to stating what we know now. He thought that matter was composed of infinitely small particles that could not be destroyed, sub-divided or compressed. This description is very similar to that for the atoms we know about nowadays.

We now have evidence that these particles are in a continual state of movement. The idea that these particles are continually moving is called the kinetic theory. The kinetic theory is used to explain (and in some cases predict) what we notice about matter. The main ideas in the kinetic theory are:

a matter is made up of tiny particles;

b the particles are moving all the time;

c how fast and how far the particles move depends on how much energy they have and how heavy they are.

In a solid the particles are arranged in regular patterns. The particles are attracted to each other and held in place by bonds. This means that the patricles are able to move all the time by vibrating but they are not free to move about at random.

In a liquid the particles are attracted to each other, but not as strongly as in a solid. This means that they can move about more. The volume of a liquid does not alter (unless it is changed by heating, for example) but the shape of the liquid changes to fit that of the container it is put in.

In a gas the particles are moving rapidly all the time, in all directions. A gas will quickly fill any container that it is put in. Particles of hydrogen have been estimated to move as fast as two kilometres per second!

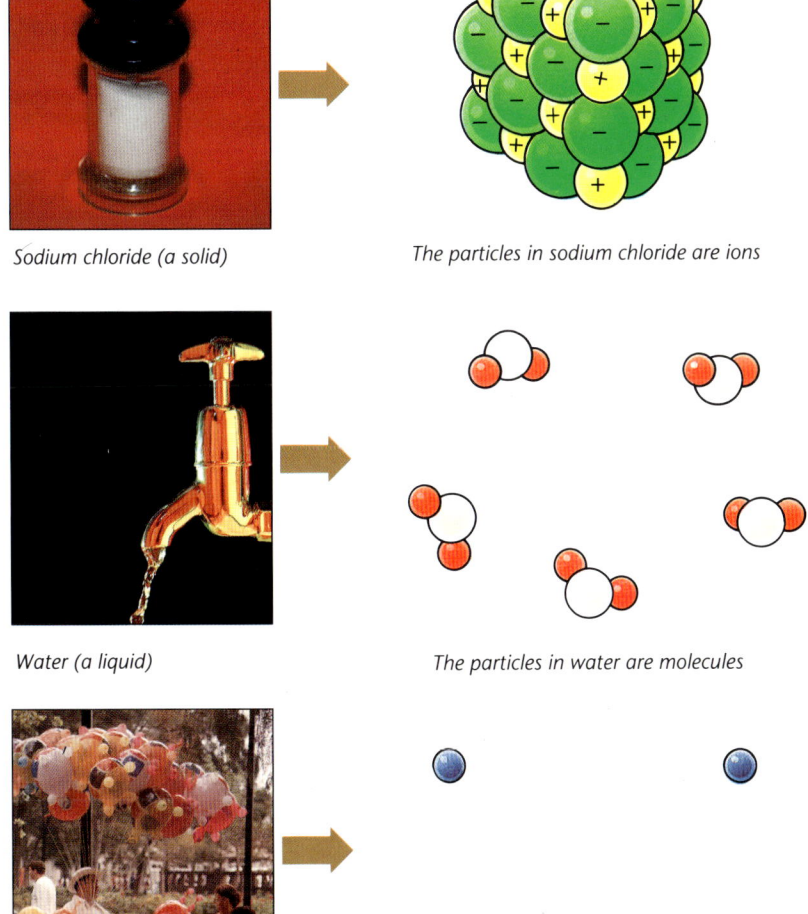

Sodium chloride (a solid)

The particles in sodium chloride are ions

Water (a liquid)

The particles in water are molecules

Helium (a gas)

The particles in helium are atoms

The kinetic theory can be used to explain phenomena such as melting, freezing, condensation, evaporation, dissolving and diffusion.

C. Design and make

- Make models that show some of the phenomena described in the previous paragraph. Each model should show the particles clearly and provide an explanation of what is happening. Add other details that you think will be helpful. Keep a note of the limitations of your model.

- Arrange the models in a display for parents, first-year pupils or some other group. Design your display especially for your target group.

  Will you need written instructions or explanations? If so, provide them.

- *Either* discuss each others models, *or* arrange for them to be evaluated by someone who was not in the class. It would be helpful if you provided a simple evaluation sheet, so that you can get some feedback about how useful your model is.

When you take an ice cube out of the freezer and put it into a drink, the drink begins to warm up the ice cube. The molecules in the ice cube get more and more energy and vibrate faster and faster. Gradually the attractions between the molecules are broken and the molecules move away from their fixed positions. The ice cube melts.

The reverse happens when you put water into the freezer to make fresh ice cubes. The water molecules give out energy as they come closer together, attractions occur between them and the water freezes.

D. Think about

- Complete the table on Worksheet PB2 to show the energy changes associated with the different changes of state.

IT'S A GAS

What is a gas? It is not easy to give an answer to this question. Scientists knew a lot about solids and liquids, before they had any clear ideas about gases. However, you should have some ideas about what a gas is.

E. Discuss

- As a group discuss the following questions:

  1 What do you know about a gas? (Be careful, your description must fit all gases!)

  2 How many gases do you know? List something special about each one.

  3 If you research the history of gases you will find a lot of references to glass apparatus. Discoveries about gases were closely linked with the development of the skills of handling glass. Explain why this was.

Robert Boyle (1627 – 1691)
Boyle thought that there was a state of matter in which the parts were separate and free to move, so that the state had no fixed shape. He collected hydrogen gas, but thought that it was a 'different modified form of air'. He thought that all substances were made up of the same 'fundamental particles'

Jan Baptist van Helmont (1580 – 1644)
Van Helmont was the first man to use the word 'gas' for substances that were like air. The word is thought to have come from the Flemish word for chaos. He was the first person to distinguish gases from air. Van Helmont first used the word 'gas' to mean a form of water. He thought that any matter when carried into the extreme upper air was turned into gas by the sharp cold and 'death of the ferments'. He said the gas might fall as rain. He did distinguish between nitric oxide, sulphur dioxide and methane but thought they were all mostly water

Sir Isaac Newton (1642 – 1727)
Newton thought that gases were made of atoms that were more or less stationary. He believed that the particles repelled each other with a force that varied inversely with the distance between them

F. Find out

- The three scientists in the photographs made some important contributions to our knowledge of gases. Record in a table, similar to the one shown here, what their main contributions were. Also record those aspects that were confusing and we now know to be inaccurate.

Name	Positive Contribution	Confusing Aspects

- Find out how the following people contributed to our knowledge of gases and add this information to your table:

Black   Scheele   Cavendish   Priestly   Lavoisier   Bernoulli

Gases are involved in many chemical reactions. When we work in a laboratory it is very important that we know how much of a substance is needed for a chemical reaction and also how much of each product is made.

It is even more important to manufacturers. They need to know how much of the reactants to buy and how big the containers for the products need to be, especially if one of the products is a gas. A large container may be very expensive. If only a small amount of gas is made then the cost of the container may be more than the value of the gas. The gas may be corrosive so the material to make the container will need to be carefully selected. A special polymer may be used that does not corrode. These polymers are expensive, so it may be decided to use the polymer just to line a cheaper metal container. These are just some of the important factors that need to be taken into consideration when deciding if a process is economical and will provide a profit for the manufacturer.

G. Plan and test

- In the reaction of magnesium with hydrochloric acid, magnesium chloride and hydrogen are made. Carry out an investigation to find out what the pattern is between the amount of magnesium used and the volume of hydrogen gas produced. Your teacher will tell you the smallest and greatest amounts of magnesium that you can use.

- Plan the investigation carefully. These are some of the questions that you will need to consider:

  How many separate experiments will you do?

  What is the best way to measure the mass of the reactants?

  What is the best way to measure the volume of hydrogen produced?

  How will you know when the reaction is over?

  What is a good way to record your results?

  How will you present your results?

- Do a trial run and note any difficulties. Revise your plan so that you overcome them.

- Carry out your investigation.

- After the investigation present your results. From your results you should be able to:

  a   show the volume of gas produced from each amount of magnesium that you used;

  b   predict the volume of hydrogen formed from amounts of magnesium less and greater than you used;

  c   say how much hydrogen was produced from an amount of magnesium between any two amounts that you used.

- Include a note on where the investigation was inaccurate and how you could improve it.

- Write down any general conclusions that you can draw from your investigation.

- Suggest other similar experiments that you could do to find out if your conclusions are true for other reactions.

HOW GASES BEHAVE

It is easy to work out the volume of a solid or a liquid. All you need is a ruler or a measuring cylinder of a suitable size and access to some water. It is not so easy to work out the volume of a gas. Imagine that you have a container of gas, such as a syringe or a balloon. Its volume can easily be changed. Think how you would do this.

The volume of a gas is altered by both temperature and pressure. These are called variables. In Section 1.2 the volume of the hydrogen gas was measured at room temperature and pressure. It is very important that the temperature and pressure of a gas are always specified because altering either of them can affect the volume of the gas.

The volume of a gas can be changed easily because gases are made up of millions and millions of tiny particles. These particles are either molecules or atoms (depending on what the gas is). Air contains millions of oxygen molecules, millions of argon atoms and many more particles. Do you remember the names of some more of the gases in the air?

While you are reading this the atoms and molecules in the air are knocking into you all the time. You do not notice them because the particles are very light. The particles in gases are moving rapidly all the time. On average they are moving faster than a plane in supersonic flight!

Changing the temperature and pressure of a gas affects how the particles move. In the following activities you will make some observations of the temperature, pressure and volume of gases and try to work out what is happening. In activity H you will investigate how pressure affects the volume of a gas. In activity I you will investigate how temperature affects the volume of a gas.

H. Work out

1 Gently press in the plunger of a sealed syringe.

What do you feel?

What happens to the volume of the air in the syringe?

Make up a simple rule that links the pressure and the volume of the air.

2 The following table show the volume of a balloon as it is submerged under water.

Depth (m)	Pressure of water (bars)	Volume of balloon (cm³)
0	0	2000
200	20	1600
400	40	1200
600	60	800
800	80	400
1000	100	200

Draw a graph to show how the volume changes with the pressure.

3 Skin divers normally stay within 30 metres of the surface of the water, yet it is still very important that they breathe out as they surface. Explain why this is.

This diver is using a rotary brush to clean deposits off a valve on an underwater pipeline. Divers often need to work in deep water. The water pressure affects the air in their bodies

In 1662 Robert Boyle carried out a number of investigations on gases. From his conclusions he was able to state a law which described how the pressure and volume of a gas are linked. This was quite an achievement. Remember there was little idea of what a gas was before this time and many still thought of gases as 'spirits'. Boyle also had poor equipment which meant that he could not take accurate readings. To try and solve this problem he had to make much of his own apparatus.

Boyle said that *'if we take a fixed mass of gas at a fixed temperature then increasing the pressure decreases the volume in proportion'*.

Boyle's law

$$\text{volume} \propto \frac{1}{\text{pressure}} \qquad p \propto \frac{1}{V} \text{ or } pV \propto 1$$

This is sometimes written as $pV = \text{constant}$

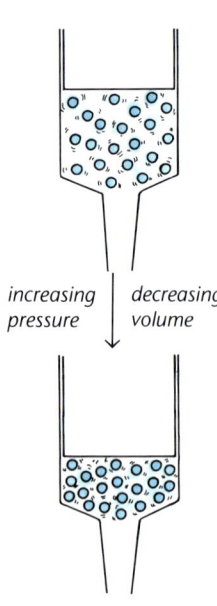

increasing pressure | *decreasing volume*

You can visualise what happens in Boyle's law if you think in terms of particles

I. Think about

1 Set up the apparatus shown in the diagram and place the lighted candle about 10 cm below one of the plastic bags.

What do you notice?

What has happened to the volume of the gas in the plastic bag.

Make up a simple rule that links temperature to volume.

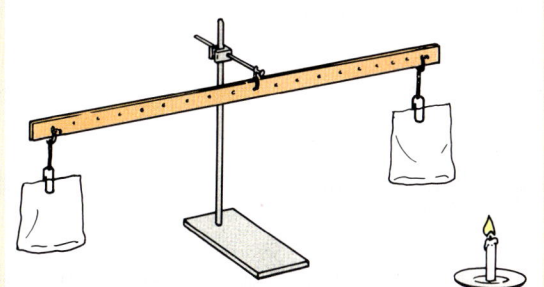

2 Put a dented ping-pong ball into a beaker of hot water and leave it there for a few minutes.

What happens?

Use your rule to explain this.

3 A sealed barrel of a gas syringe was connected to a tube. The syringe contained some oil and the tube, also full of oil, had an overflow reservoir. The apparatus was set up as in the diagram.

The volume of the gas in the syringe was taken at intervals as the water was heated.

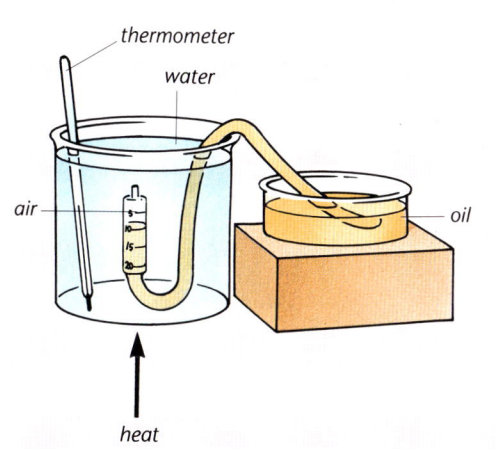

These are the results:

Temperature (°C)	Volume (cm³)
5	10.65
25	11.5
50	12.4
75	13.4
100	14.4

Draw a graph to show how the volume changes with the temperature. (Make sure that the temperature goes from -300°C to 110°C and the volume from 0 cm³ to 15 cm³.)

It was in 1787 that Jacques-Alexandre Charles collected together the results of a number of investigations into how temperature affects the volume of a gas. He finally overcame the biggest problem which was that he had inaccurate thermometers. He came up with a law which we now know as Charles' law.

He said that *'if we take a fixed mass of gas and do not alter the pressure then as the temperature of the gas increases the volume increases in proportion'*. In other words the volume is proportional to the temperature.

Charles' law

$V \propto T$

This is sometimes written as

$V = \text{constant} \times T$, or $\dfrac{V}{T} = \text{constant}$

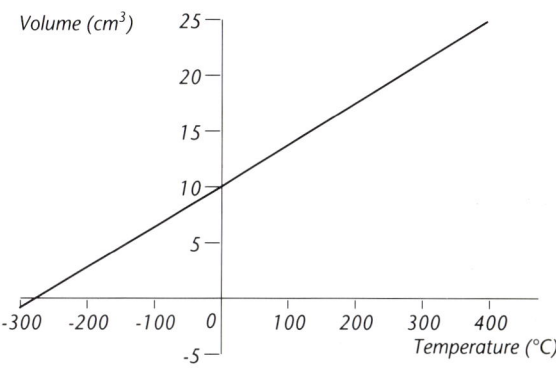

Graph showing how the temperature affects the volume of a gas

Look at your graph for the gas syringe experiment in the previous activity. If the graph is extrapolated back you will see that the volume would be zero at -273°C.

This is the temperature at which particles of all substances stop moving. This suggests that -273°C is the coldest that anything can be! This temperature is called absolute zero. Scientists often use kelvins to measure temperature. The Kelvin scale begins at absolute zero and increases just like the Celsius scale: 0°C is 273K, room temperature is about 20°C or 293K and the boiling point of water is 100°C or 373K. (It is easy to turn °Celsius into kelvins – you just add 273!)

In the next activity you will investigate how pressure affects the temperature of a gas.

J. Interpret

1 Find out what the pressure of a car tyre is before and after a journey.

What has happened to the temperature of the tyre during the journey?

What has happened to the pressure of the air in the tyre?

Make up a simple rule that links temperature and pressure.

2 Pamela and Chris carried out an investigation in which they heated air and took its pressure at regular intervals. A record of what they did and the results they obtained are shown here.

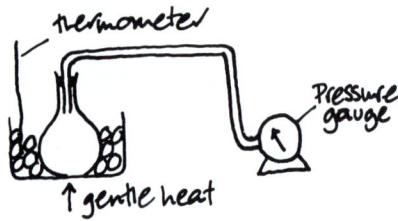

To find out how temperature affects the pressure of the air.

We set our experiment up like this:

(diagram: thermometer, pressure gauge, gentle heat)

We took the pressure with the ice around the flask. Then, we gently heated it for quite a while — until the ice had turned into water and boiled.

Our results are:

Temperature /°C	Pressure N/m²
0	100 000
25	110 000
50	118 000
75	127 000
100	136 000

Convert the temperatures to kelvins and plot their results for them. Start at 0 kelvin.

Does their experiment confirm your ideas about the link between the temperature of air and its pressure?

The pressure law describes how temperature and pressure are linked:

For a given mass of gas, if the volume is kept constant then the pressure of the gas is directly proportional to the temperature. (Remember that kelvins are used as the unit for temperature.)

The pressure law

$p \propto T$

This can be written as: $p = \text{constant} \times T$

or $\dfrac{p}{T} = \text{constant}$

Temperature is important in regulating the pressure of the air in a hot air balloon

K. Work out

• Make up your own summary about the three laws that help us to understand more about what gases do. For each of the laws try and include:

a the equation;

b a sketch graph to show how the two quantities are related;

c information about which quantities must be kept constant for the law to be true;

d any special rule (for example, when kelvins must be used);

e an explanation about what happens to the particles;

f some real examples of the law.

• Make a leaflet which contains inside it the facts about how gases behave. On the outside write some questions about gases. Make a 'rough' leaflet first, and ask a friend to judge it; improve it if necessary. Make your final leaflet as interesting as you can. You will be able to use this leaflet later for revision.

Extension

Earlier in the section you found out how pressure and temperature affect the volume of a gas. In each of the activities you investigated two of the variables at a time while you kept the other one constant. You ended up with these three equations:

Boyle's law

pV = constant (Boyle's law constant)

Charles' law

$\dfrac{V}{T}$ = constant (Charles' law constant)

The pressure law

$\dfrac{p}{T}$ = constant (pressure law constant)

The laws can be mathematically combined to give one equation:

$\dfrac{pV}{T}$ = constant

This is called the ideal gas equation. It can be used to convert from one set of conditions to another.

The ideal gas equation

$$\dfrac{p_1 V_1}{T_1} = \dfrac{p_2 V_2}{T_2}$$

where p_1, V_1 and T_1 are the pressure, volume and temperature of a gas at one time and p_2, V_2 and T_2 represent what the gas is like later.

You are probably wondering what an ideal gas is. The characteristics of an ideal gas are:

a The particles only move in straight lines.

b The particles move about in a random way. They only change direction when they hit other particles or the walls of the container.

c When the particles collide they do not lose any energy.

d The particles do not attract or repel each other.

e The volume of the particles can be ignored.

In reality, there is no such thing as an ideal gas but the gases we know (called real gases) fit this law well enough for it to be useful.

Example

Sulphur dioxide gas is formed in a reaction. At a temperature of 0°C and a pressure of 760 mm Hg the volume produced is 1400 cm³. The reaction is carried out at a pressure of 850 mm Hg and a temperature of 250°C. What volume of sulphur dioxide is produced?
(1 mm Hg = 133.3 Pa)

First convert the temperature to kelvins:
initial temperature = 273K
final temperature = 523K

Now write down the conditions like this.

p_1 = 760 mm Hg	p_2 = 850 mm Hg
V_1 = 1400 cm³	V_2 = ?
T_1 = 273K	T_2 = 523K

Substitute these values in the ideal gas equation:

$$\dfrac{p_1 \times V_1}{T_1} = \dfrac{p_2 \times V_2}{T_2}$$

$$\dfrac{760 \times 1400}{273} = \dfrac{850 \times V_2}{523}$$

$$V_2 = \dfrac{760 \times 1400 \times 523}{273 \times 850} = 2398 \text{ cm}^3$$

The volume at 850 mm Hg and 250°C will be 2398 cm³.

L. Work out　　　　　　　**EXTENSION**

1 350 cm³ of air in a diesel engine at a temperature of 25°C and pressure of 750 mm Hg are compressed by the piston. The piston raises the pressure to 30 000 mm Hg and the volume is reduced to 250 cm³. By how much does the temperature rise?

2 The volume of a gas was measured at 35°C and 750 mm Hg pressure and was found to be 60 cm³. What would the volume be at 0°C and 760 mm Hg?

3 If a factory is producing 8 million litres of ammonia at 200 atm and 400°C, what volume of ammonia is produced if it is released as a gas at room temperature (25°C) and a pressure of 1 atmosphere?

GAS CONTROL

The particles in a gas are spread out and they are moving about all the time. It is not difficult to make the particles spread out even more or to squash them more closely together.

M. Work out

- The illustration shows several examples of the pressure of a gas being changed. Work out what is happening to the pressure, the volume and the particles in the gas for at least three of these examples. It may be useful to present your answers in a table similar to this one:

Example	Pressure change	Volume change	Diagrams showing how the particles in the gas change	Notes

---

The manufacture of many products involves gases. The effect of temperature and pressure on these gases is very important during the manufacturing process. Ammonia is one of the most important chemical substances being manufactured today. It is estimated that 5500 tonnes are made each day in this country. That is enough to fill 8000 million empty litre bottles.

In *Chemical processes* Unit 6 you saw how Fritz Haber managed to get an 8% yield of ammonia in the following reaction:

$$N_2(g) + 3H_2(g) \rightleftharpoons 2NH_3(g) \qquad \Delta H = -92 \text{ kJ mol}^{-1}$$

This reaction reaches an equilibrium. The word 'equilibrium' is used to describe a reversible reaction in which the reactants are turning back into products at the same speed as the products are turning into reactants. By changing the conditions Haber could increase the amount of product. He used a principle which simply stated says:

When you alter the conditions of a reaction that has reached an equilibrium, the reaction will change in such a way as to reduce the effect of what you have done.

Haber used the equation and this principle to give him two vital pieces of information:

1. One mole of nitrogen combines with three moles of hydrogen to form two moles of ammonia. In other words, four moles of reactants form only two moles of ammonia. Using the principle, he knew that if the pressure was increased then the reaction would change to try and make the pressure less. This could be done by turning more of the four moles of reactants into two moles of product. More ammonia could be produced!

2. As ammonia is produced heat is given out; the reaction making ammonia is exothermic. By using the principle again, Haber predicted that by decreasing the temperature the reaction would change to try to increase the temperature. It could do this by making more ammonia.

The graphs show how the temperature and pressure affect the yield of ammonia.

For the process of making ammonia to be economic there are a number of factors that need to be considered. Two of the most important ones are:

a   yield (how much ammonia is produced);

b   rate (how fast it is produced).

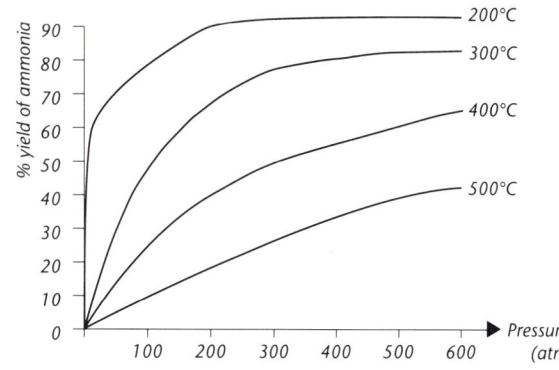

The yield of ammonia at different temperatures and pressures

N. Think about

- Imagine that you have been called in to give advice on the production of ammonia. You will need to decide on the ideal conditions for the reaction. Some points you will need to think about are:

  What is the best temperature for maximum yield?

  How will you achieve this temperature?

  What is the best pressure for the maximum yield?

  How will you achieve this pressure?

  What problems will your chosen pressure give you, say if there is a leak in a piece of piping?

  How will the chosen temperature and pressure affect the rate of the reaction?

- A catalyst can be used to increase the rate of the reaction. The graph shows the effect that some catalysts have on the production of ammonia.

  Which catalyst would you recommend to improve the rate of reaction?

- Many manufacturing processes involve competing priorities. Using ammonia production as an example:

  a   state some of the competing priorities;

  b   write a report of your suggestions for resolving them.

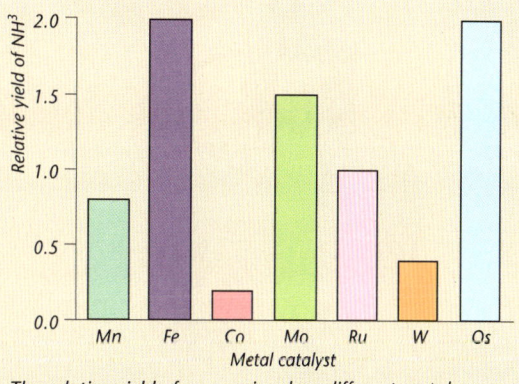

The relative yield of ammonia when different metal catalysts are used

---

Blasting rock for a new airport. Ammonia is used in the production of explosives

In practise the process of producing ammonia is always a compromise. It is necessary to balance the yield with a sufficiently fast rate of reaction. It is just not economic to have a very slow process. The following conditions are used:

pressure   200 atmospheres

temperature   380–450°C

catalyst   iron (with promoters to stop it being poisoned by the ammonia)

yield   15% by volume of ammonia

These conditions enable the production of the ammonia to be as economic as possible.

When a new plant to produce ammonia is developed, computers are used to simulate or model the conditions which could arise. All manner of problems can be tried out in the computer model. In this way an efficient and reliable plant can be built. This is much more economical than modifying or improving a plant that is already built.

Storing large quantities of gas can be quite a problem, due to the space needed. One way round the problem is to turn the gas into a liquid which would take up less space. How this could be done was a question that occupied Faraday between 1823 and 1845. He systematically tried to liquefy all known gases. He tried compression and cooling. It was not until 1895 that air was finally liquefied. Today compressed and liquefied gases have many uses.

O. Work out

- Name two of the compressed or liquefied gases in the illustration.

- Give a use for each of the gases.

- Explain how the gas has been turned into its present state.

  What are the advantages of the gas being in this state?

  What are the disadvantages (if any) of this?

Aerosols

Aerosols are widely used today. Early 'aerosols' produced a fine spray of mist which stayed in the air, a sort of suspension, or solution, of liquid in air. For this reason the pressurised packages became known as aerosols (aero = air, sol = solution). This word is actually used in many languages. The label on an aerosol indicates that it is a pressurised container.

P. Research

- Use Worksheet PB5 *Finding out about aerosols* to help you to find out how much people know about aerosols.

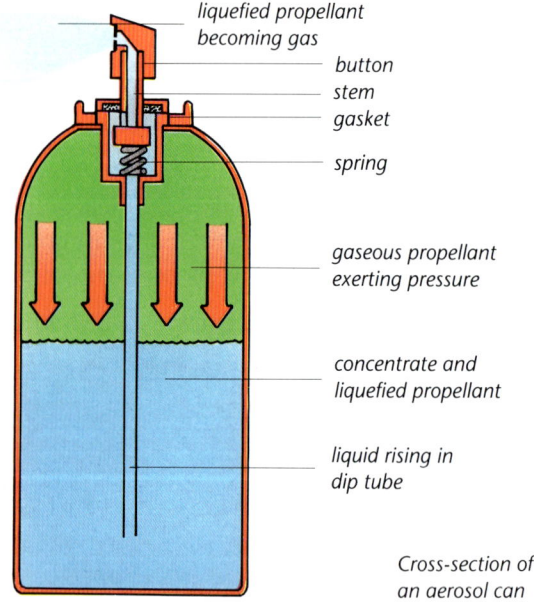

liquefied propellant becoming gas

button
stem
gasket

spring

gaseous propellant exerting pressure

concentrate and liquefied propellant

liquid rising in dip tube

Cross-section of an aerosol can

Aerosols are widely used

UNIT 2
THE HEART OF THE MATTER

Section 2.1

INSIDE THE ATOM

One of the biggest steps in understanding the nature of matter was made by John Dalton around 1800. Describing one of his most important discoveries he said *'Every particle of water is like every other particle of water; every particle of hydrogen is like every other particle of hydrogen.'*

We now know much more about these particles that Dalton described. From earlier work you will remember that atoms and molecules are two types of particles.

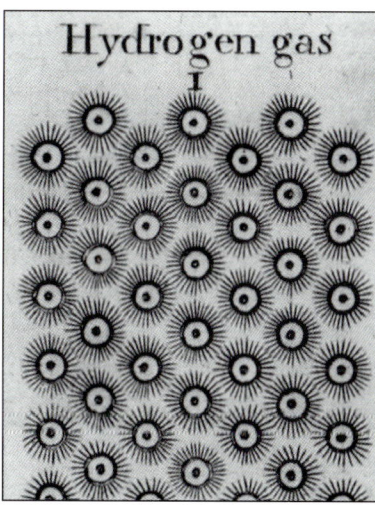

John Dalton's diagrams of hydrogen and carbon dioxide (carbonic acid gas)

A. Discuss

- First, write down what you yourself know about:

  a an atom;

  b a molecule.

- Then, as a group, try and agree on a list of facts about each of these types of particle. You will need to look up things that you are not sure about. Some questions that might help you with your discussion are:

  What is the difference between an atom and a molecule?

  What are the particles in water?

  Are there atoms in water, if so what are they?

  What are the atoms and molecules in the air?

  How many different kinds of atom are there?

  How many different kinds of molecule are there?

- Record what you decide on or find out under the headings **Facts about atoms** and **Facts about molecules**. It may help to draw some diagrams.

- The photographs show John Dalton's diagrams of hydrogen and carbon dioxide (he called it carbonic acid gas). Draw one or two of the particles of each of these gases.

  How is Dalton's diagram of hydrogen inaccurate?

- Write the (correct) chemical shorthand for the gases hydrogen and carbon dioxide.

The carbon atoms in the carbon dioxide molecules that you have just breathed out have been around since the Earth was formed. They may have been part of the first atmosphere which collected around the Earth. Where have these atoms been? From earlier work you will know that carbon atoms move round in a complex cycle. Who knows, they may have been breathed out by Tutankhamen or even been to the Moon and back on one of the many space missions. It is difficult to believe, but remember that atoms can not be created or destroyed so the same atoms must always have been around.

Atoms are amazingly small. Most have a radius of about 10^{-10} metres. This is very difficult to imagine, but it may be easier if you realise that each page in this book is about two million atoms thick. Individual atoms can now be seen as light spots using an electron microscope.

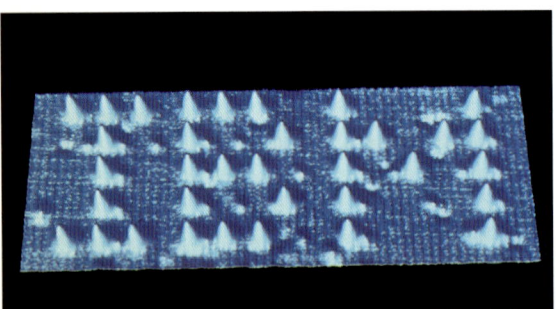

These IBM letters are about 500 000 times smaller than those on this page. The distance between atoms in the finished pattern is about 50 billionths of an inch (13 Angstroms). The entire IBM is about 660 billionths of an inch (168 Angstroms) in length

Dalton thought that the smallest part of an element was an atom, but we now know that there are smaller particles making up the atom itself. These are known as sub-atomic particles. Our knowledge of these particles and how they are arranged in the atom is based on indirect evidence. You will have met some of this evidence already.

Sometimes you will have noticed a crackling noise when you have taken nylon clothes off. This is an example of the formation of an electric charge called static electricity. There are two sorts of charge – positive and negative. Ordinary atoms have equal amounts of both sorts of charge. They are electrically neutral.

Negatively charged sub-atomic particles are called electrons

Electrons were the first sub-atomic particles to be discovered. They were named by J J Thomson in 1897.

Electrons are known to be negatively charged and very light. Nearly 2000 electrons have the same mass as one atom of hydrogen. We can picture the electrons as moving around the outer part of the atom. This means that when atoms react in a chemical reaction it is the electrons that are involved. A previous model suggested that electrons orbited the nucleus like planets. A better model is that the electrons are arranged in 'shells'. A shell is not a fixed position but is related to how much energy an electron has.

Positively charged sub-atomic particles are called protons

The protons of an atom are found in a small central area called the nucleus. Each proton weighs about the same as a hydrogen atom (called 1 amu or atomic mass unit). In an atom there are as many positively-charged protons as there are negatively-charged electrons. That is what makes the atom have a neutral electrical charge overall.

Using this apparatus J J Thomson identified cathode rays as streams of negatively-charged particles which he called electrons. The principle of the apparatus is the same as that for a TV where an electron gun fires a beam of electrons at the screen

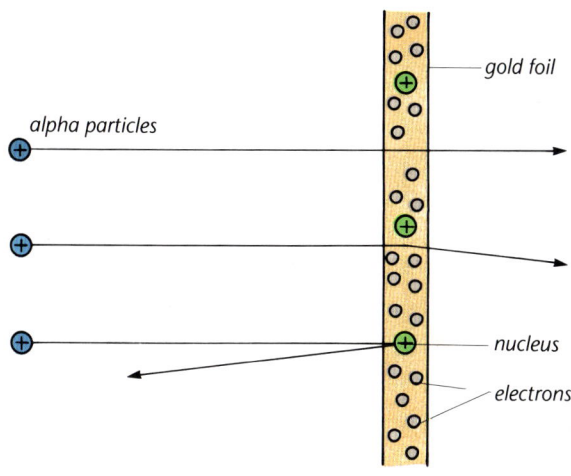

Two experimenters, Geiger and Marsden, fired a stream of positive particles at a thin sheet of gold. Most of the particles went straight through, but some were deflected and others rebounded. From this Rutherford suggested that atoms must contain a small positively-charged nucleus. He was able later to propose the existence of protons inside the nucleus

Neutral sub-atomic particles are called neutrons

Besides containing protons the nucleus contains other particles called neutrons. Each of these weighs 1 amu, but it has no charge.

In 1932 Sir James Chadwick, using this apparatus, bombarded the element beryllium with positive particles. The beryllium gave out particles which were neither positively or negatively charged. Since these particles were neutral they were called neutrons

All atoms are built up of these three basic particles: electrons, protons and neutrons. The atoms of the different elements have different masses because they contain different numbers of protons and neutrons in their atoms. In most of the periodic tables which show the elements you will see some numbers in front of the symbol for each element. For example, $_{11}^{23}$Na.

1 The Na tells us that it is a sodium atom.

2 The 11 tells us that it has 11 protons. This is called the atomic number (symbol Z).

3 The 23 tells us that it has a total of 23 protons and neutrons in the nucleus. This is called the mass number and is given the symbol A.

mass number ↘

$$_{11}^{23}\text{Na} \leftarrow \text{indicates it is sodium}$$

atomic number ↗

We can also work out that an atom of sodium has:

4 11 electrons (in an atom there are the same number of protons and electrons).

5 12 neutrons (if an atom has 11 protons, it must have 12 neutrons to give a mass number of 23).

B. Work out

- Copy and complete this summary table to compare the particles in an atom:

Name of particle	Where it is found in the atom	Charge	Mass (in amu)
electron			
neutron			
proton			

- Copy out the following table and complete it for these elements:

  $_{1}^{1}$H, $_{3}^{7}$Li, $_{10}^{20}$Ne, and $_{13}^{27}$Al

Symbol	Name	Number of protons	Number of electrons	Number of neutrons

ELECTRON DISTRIBUTION

If we magnified an atom one million million times to the size of Wembley stadium, the nucleus would be the size of a small marble placed at the centre of the pitch. Spectators at the back of the stand would have the outer electrons moving behind their heads

Most of an atom is in fact empty space. The nucleus of neutrons and protons is tightly packed together. The electrons surround this nucleus but keep a long way from it. The electrons do not move randomly throughout the space around the nucleus. Niels Bohr (1885 –1962) thought the electrons were arranged in shells, rather like the layers of an onion. By taking measurements of how much energy it takes to remove an electron from an atom the idea of electrons being in shells is confirmed.

The sodium atom has 11 electrons and the energy needed to remove each electron is given in the following table:

N.B. The first electron is *furthest* from the nucleus

Number of electron	Energy needed to remove each electron (kJ mol^{-1})
1st	496
2nd	4563
3rd	6913
4th	9544
5th	13352
6th	16611
7th	20115
8th	25491
9th	28934
10th	141367
11th	159079

You can see from the table that it does not take much energy to remove the first electron. Then it gets progressively harder until the ninth one is removed. It is very much harder to remove the last two electrons. This suggests that the outermost electron is in a shell by itself, the next eight are in a shell together, and the last two are in an innermost shell. This confirms Bohr's 'electron shell' idea. Bohr's electron shell theory simply says that:

1 Electrons are arranged in shells around the nucleus.

2 The first shell, nearest the nucleus, can only contain up to 2 electrons.

   The second shell can contain up to 8 electrons.

   The third shell can contain up to 18 electrons. However, for small atoms (those with up to twenty electrons) the third shell will not hold more than eight.

3 The electron shell nearest to the nucleus (which only contains two electrons) fills first.

4 The outer electrons can be removed fairly easily but it gets progressively harder to remove the electrons as you move towards the nucleus.

Using Bohr's theory, a sodium atom can be represented as:

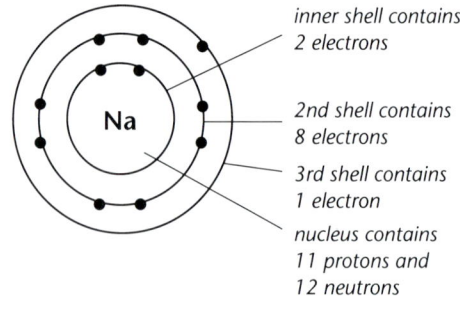

inner shell contains 2 electrons

2nd shell contains 8 electrons

3rd shell contains 1 electron

nucleus contains 11 protons and 12 neutrons

It can also be written as Na 2,8,1 which shows how the electrons are arranged in layers.

C. Design and make

- Choose one of the first twenty elements from the periodic table and make a model to show its atomic structure. When designing the model you will need to decide:

  a   Will it be two-dimensional or three-dimensional?

  b   How will you indicate the relative sizes of the particles and their charges?

  c   How will you encourage your audience to take an interest in your model?

  d   Should you label it?

  e   Should you mention the uses of the element which the model represents an atom of?

- Keep notes during the construction of the model. You will need to evaluate it when you have finished.

- Finally, prepare an instruction sheet, so that the model can be made again by someone else. Include notes to say how the model could be altered to represent any one of the first 20 elements.

- As an appendix to your instruction sheet compare your model with your mental picture of a real atom.

What are ions?

Earlier in this section you saw how much energy was needed to remove each of the electrons, in turn, from a sodium atom. The sodium atom is electrically neutral, that is, it has as many positive protons as it does negative electrons. If a negative electron is taken away from it there will be 11 protons in the nucleus but only 10 electrons. So, overall it will have a positive charge and so cannot be called an atom. It is now called an ion. This is written as Na^+ (to show that the charge is 1+).

We can write the change like this:

$Na \quad - e^- \quad \rightarrow Na^+ \qquad$ **or** $Na \rightarrow Na^+ + e^-$
sodium   loses an   to form a
atom   electron   sodium ion

Similarly an atom can gain an electron. For example, a chlorine atom has 17 electrons and by gaining another it will have 18. It now has more electrons than protons and so has a negative charge.

We write this as Cl^- (to show the charge is 1-).

We can write the change like this:

$Cl \qquad + \quad e^- \qquad \rightarrow \quad Cl^-$
chlorine   gains an   to form a negative
atom   electron   chloride ion

What is an isotope?
If you have a smoke detector at home it probably contains a radioactive isotope. But, what is an isotope?

Hydrogen has three isotopes. The type that we met earlier in this section is ^1_1H.

The other two kinds of hydrogen atom are ^2_1H, which is called deuterium, and ^3_1H, which is called tritium.

You will see that all three kinds of hydrogen atom have the same number of protons (and therefore electrons) but different numbers of neutrons. Isotopes of an element have the same atomic number (Z) but different mass numbers (A). Of the three isotopes of hydrogen only tritium is radioactive. You will find out more about isotopes in Unit 4 of this topic.

A radioactive isotope can be found in a smoke detector

D. Interpret   **EXTENSION**

- Using chemical shorthand it is possible to give a lot of information about an atom or particle. For each of the following work out as many facts about it as you can.

  1   F 2,7   6 $^{12}_6C$, $^{13}_6C$, $^{14}_6C$

  2   Ca^{2+} 7   potassium 2,8,8,1

  3   ^4_2He 8 Al^{3+}

  4   S_8 9 NO_3^-

  5   Br^- 10 $^{257}_{103}Lw$

ATOMIC BONDING

If you look at the periodic table you can find out how many known elements there are. It is not so easy to say how many compounds there are. New ones are being made all the time.

In the last section you considered the structure of the atom and saw that the electrons are on the outside of the atom. It is these electrons that are the most likely to be involved when atoms react to form a new molecule. At first it may seem to you that it is unlikely that atoms ever react. When two atoms come together you would expect their outer electrons to repel each other and to force the atoms to move apart, rather than to react together. However, when two atoms come together, one of three things can happen:

- the two atoms move away again and no bond is formed between them;

- one atom 'gives' electrons to the other atom and a bond is formed between them;

- the two atoms share some electrons and a bond is formed between them.

The two atoms move away and no bond is formed

The air contains a number of different types of atom, for example one percent of the air is made up of argon atoms. There are millions of argon atoms moving around the room at this minute and the chances are that some will collide with each other. However these atoms will move apart and they will not bond together. By looking at the electron structure of argon, Ar 2,8,8, we can get a clue as to why this is. The outer electron shell is 'full'; it contains eight electrons. This is an important fact. When an atom has a full outer shell it seems unwilling to accept or lose electrons, or even to share them. It is unreactive.

Tip for recognising elements that have unreactive atoms

- The outer shell of electrons is 'full'.

Bond formed when one atom 'gives' electrons to another atom

The electron structure of sodium is Na 2,8,1 and chlorine, Cl 2,8,7. If we use the evidence that we have just seen about argon – that the atoms are stable when they have a full outer shell of electrons – we can suggest what will happen in this case. The sodium atom has one extra electron beyond what it needs for two full shells. The chlorine atom is one short of having three full shells. If the sodium atom 'gives' an electron to the chlorine atom, then both the atoms will have full shells.

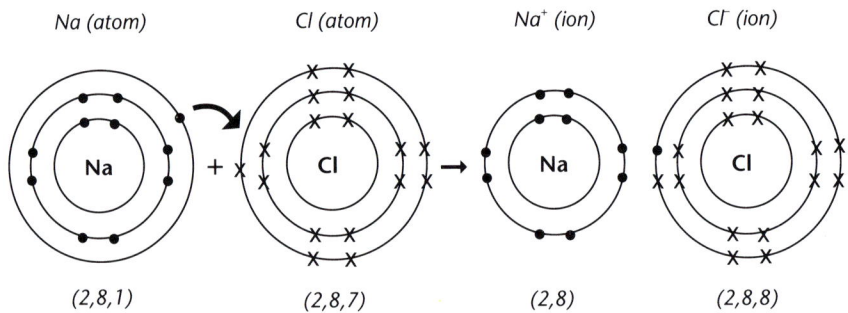

Na (atom)   Cl (atom)   Na$^+$ (ion)   Cl$^-$ (ion)

(2,8,1)   (2,8,7)   (2,8)   (2,8,8)

(The sodium electrons are shown as dots and the chlorine electrons as crosses to make it easier for you to see what is happening)

The sodium and chlorine ions formed by this process have opposite charges so they attract each other strongly. They are held together by an ionic bond and this can be written as Na$^+$Cl$^-$.

Tips for recognising substances which have molecules with an ionic bond:

- they have high melting and boiling points;

- they form crystals;

- they conduct electricity when in aqueous solution and when molten;

- often they are soluble in water;

- one of the elements is a metal.

Bond formed when two atoms share electrons

Argon exists as individual atoms but many gases, such as chlorine (Cl_2), exist as molecules. By looking at the electron structure of chlorine, we can suggest what happens to the outer electrons. Chlorine atoms are one short of the eight which will make them have a full shell of outer electrons. However, if two chlorine atoms share a pair of electrons, one from each of them, they can both have a full outer shell of eight electrons.

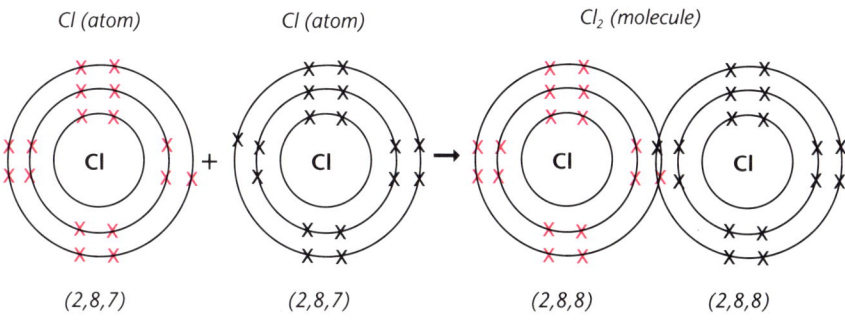

Cl (atom)　　　Cl (atom)　　　　　　　Cl_2 (molecule)

(2,8,7)　　　(2,8,7)　　　(2,8,8)　　　(2,8,8)

(The electrons from the two atoms are shown in different colours to make it easier for you to see what is happening. The two chlorine atoms share two electrons)

The pair of electrons that are shared make up a covalent bond and this can be written as Cl-Cl. The new particle that is formed is called a molecule.

Tips for recognising substances which have molecules with covalent bonding:

- they have low melting and boiling points;
- they involve only non-metals;
- they do not conduct electricity.

E. Investigate

- You are given four substances. The particles in some of these substances are held together by ionic bonding and in the others by covalent bonding. Your task is to sort out which is which by carrying out experiments and looking up data.

  You will need to know these features about each substance:

  a　What its appearance is like.

  b　Whether it has a smell.

  c　Whether it is soluble in water.

  d　Whether an aqueous solution conducts electricity.

  e　How it changes on heating.

  f　What its melting and boiling points are.

- Write a report in which you state clearly which type of bonding each susbstance has. You will need to provide evidence for each decision.

F. Work out   EXTENSION

- Make two columns headed **Ionic** and **Covalent** and place each of the following in the correct column:

  bonding between two non-metals

  bonding between metal and non-metal

  conducts electricity when molten

  copper(II) chloride

  sulphur dioxide

  formed by giving away and receiving electrons

  formed by sharing electrons

  gas

  hydrogen molecule

  made up of molecules.

Metals do not seem to fit into either our ionic or covalent bonding groups.

However, one of the important properties of metals is that they conduct electricity. This means that the electrons must move through the metal. This is a clue to their bonding. Metals generally only have a few electrons in their outer shells. Sodium, for example, only has one. When metal atoms come together, the outer electrons seem to be shared equally between all the nuclei; they act as a sort of 'electron glue'.

The outer electrons are spread or delocalised, but the nuclei of the atoms stay in a fixed pattern. The electrons can move through the metal which is what happens when the metal conducts electricity. This type of bonding is called metallic bonding.

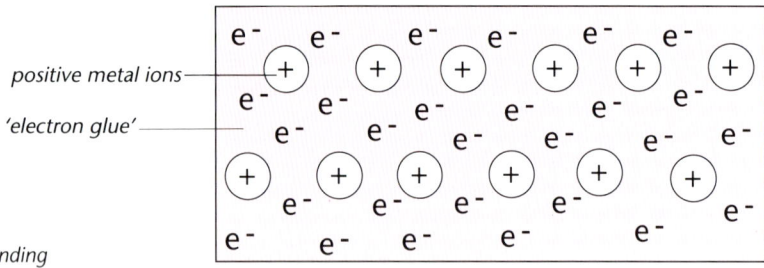

positive metal ions
'electron glue'

Metal bonding

A Californian roller coaster

Look at the roller coaster in the photograph. The overall structure and the way it runs depend on the bits it is made of, and how they are held together. Similarly with chemical substances; their structure and properties depend on the bits they are made of and how they are held together.

You have found out about the 'bits'. These are the atoms, molecules and ions. You have also looked at bonding. This can be ionic, covalent or metallic. Now you are ready to consider different ways of combining these two to see how they affect the properties and so the uses of the different substance that are made.

The information can be given in a simple key like the one shown on the next page.

Metalic bonding enables metals to conduct electricity

G. Think about

- You have been given tips for recognising ionic and covalent bonding. Write your own tips for recognising metallic bonding.

H. Work out

- Make a poster or collage that will show the structure and bonding in some everyday things. (You may concentrate on one item such as a light bulb or a car, or on a scene such as inside a building.) You should try and include the four categories that are shown in the table.

  You will need to consider these points:

  Who will see the poster? (For example, it could be aimed at adults who want to know more about science.)

  How will you make it interesting?

  How can you make it 'active' so there is something to do? (Will there be a question or puzzle with it?)

  How will you evaluate how successful your poster has been?

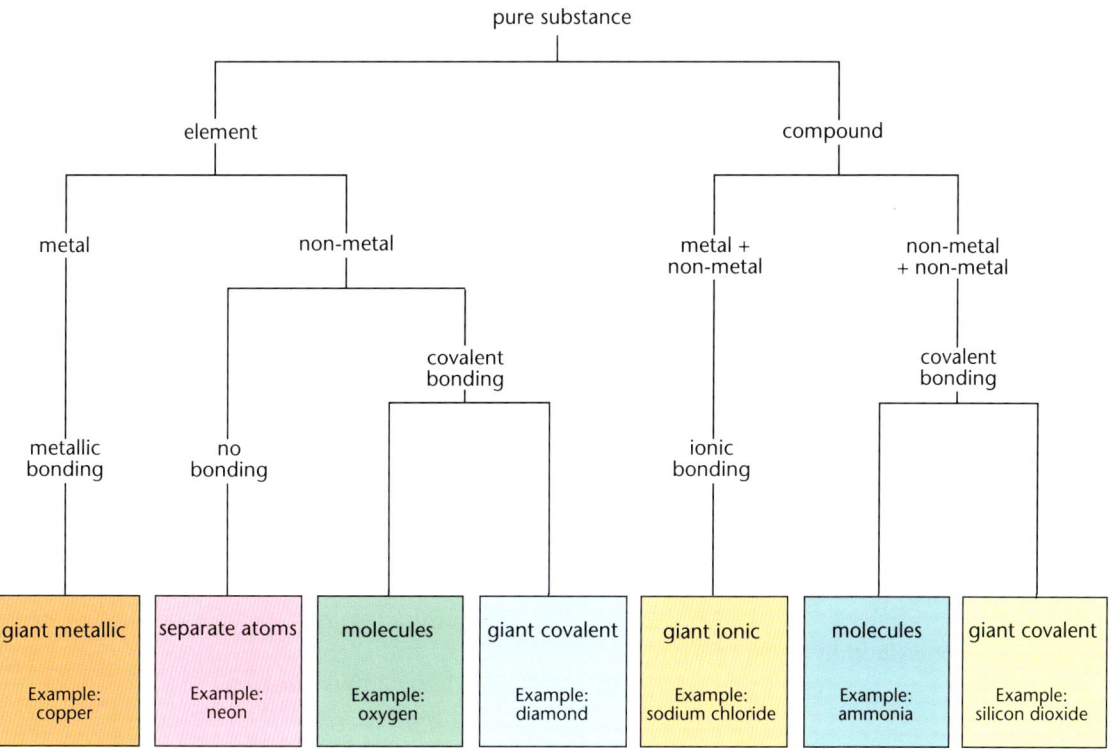

The table shows more details of each category.

Type	Examples	Structure	Typical properties
simple molecular	carbon dioxide water chlorine ammonia	Example: carbon dioxide *very weak forces between molecules* *strong covalent bonds in molecule*	• low melting point • low boiling point • usually gas or liquid at room temperature • if solid, soft and melts easily • does not conduct electricity
giant covalent	diamond silicon dioxide	Example: diamond	• high melting point • high boiling point • difficult to melt • hard but brittle • insoluble • does not conduct electricity (except graphite)
giant ionic	sodium chloride magnesium oxide	Example: sodium chloride	• high melting point • high boiling point • hard but brittle • usually soluble • conducts electricity in aqueous solution and when molten
giant metallic	copper aluminium	Example: copper	• high melting point • high boiling point • hard but malleable (can be bent and stay in shape) • most are dense • insoluble • conducts electricity

REACTION PATTERNS

Section 3.1

REACTION SPOTTER

At this very moment many chemical reactions are going on inside you. You have already considered one important one, respiration, in the topic *Chemical processes*. Respiration is a chemical reaction in which sugars are oxidised to water and carbon dioxide and energy is released.

$$C_6H_{12}O_6(s) + 6O_2(g) \rightarrow 6H_2O(l) + 6CO_2(g) \qquad \Delta H = -2816 \text{ kJ mol}^{-1}$$

sugar   oxygen   water   carbon dioxide   exothermic
(energy is released)

This reaction is one of a type which is called oxidation. As you found out in *Chemical processes* Unit 2 it is very much like the combustion process in which fuel is burnt. Combustion is another type of oxidation reaction.

$$CH_4(g) + 2O_2(g) \rightarrow 2H_2O(l) + CO_2(g) \qquad \Delta H = -890.4 \text{ kJ mol}^{-1}$$

methane   oxygen   water   carbon dioxide   exothermic
(energy is released)

Three-fifths of an oil platform is permanently under water and another fifth is constantly lashed by the sea. Corrosion here could be a very serious problem

Rusting, very much a nuisance reaction, is also an oxidation reaction. This and similar types of corrosion cost a lot of money each year.

You will be able to think of many examples of items that have to be treated to stop corrosion. Corrosion reactions are also oxidation reactions.

You have just seen how three seemingly quite different reactions, respiration, combustion and corrosion, are all really of one type, oxidation. Many reactions can be grouped into just a few different types. By finding out about these types of reaction it will help you to work out what is going on in any reaction that you come across.

Different types of reaction

Fermentation
The production of ethanol and carbon dioxide by a microbe called yeast

Polymerisation
The joining together of many small molecules to make a very large molecule indeed

Precipitation
When two solutions containing ions are mixed and produce an insoluble product

Oxidation
The addition of oxygen or removal of hydrogen to form a compound

Salt formation
The reaction in which an acid and an alkali neutralise one another

Reduction
The removal of oxygen or addition of hydrogen to a substance

Chemical reactions are important in providing us with many of the things that we need each day. Even if you do not go into a science laboratory you can still see chemical reactions going on all the time – if you look around carefully.

A. Discuss

- The illustration shows a number of chemical reactions taking place and some important everyday products produced by chemical reactions. Match each of these reactions with one of the reaction types.

Six different types of reaction have been mentioned. If these could be grouped together in some way it would be even easier to work out what is going on. One attempt at grouping them is shown here:

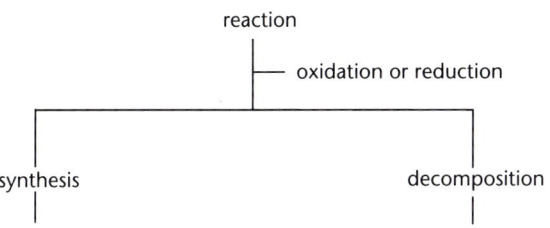

When you see a chemical reaction that you have not met before, you can make a number of predictions about it. In the following activity you will be asked to make predictions, then carry out the experiment and finally assess your predictions. You will see examples of all the different reaction types that have been mentioned.

B. Work out

- Carry out the experiments described on Worksheet PB7 *Reaction spotter*.

- Organise yourself so that you can record information about each experiment as you go along. It may help to make up a table in which you can record your predictions and results. Read through the following list of questions carefully before making up your table:

Before you begin the experiment

1   What are the reactants?

During the experiment

1   Is the product (or products) made by splitting up the reactants (called decomposition) or made by building up the reactants (called synthesis)?

2   Which of the types of reaction does it seem to fit? Give reasons.

After the experiment

1   Write a word equation for the reaction.

2   If you can, write a symbolic equation, including the state symbols.

3   Find two other reactions which fit this reaction type.

Finally

1   When you have completed as many experiments as you can, draw the reaction tree and add more branches to it to show the different types of reaction.

2   Explain any problems that you have in trying to do this.

C. Work out

- Choose one of the six types of reaction shown on page 78. Make a poster linking this type of reaction to something that you see each day or to a manufacturing process. Plan the poster carefully – you must make it eye-catching. Try and include a clear diagram or even some moveable parts to show what is going on.

MORE ABOUT WRITING EQUATIONS

Many of the reactions you have been looking at involve ions. You have already learnt how to write word and balanced equations but when the reaction involves ions it is more accurate to use ionic equations to describe it. For example, in a neutralisation reaction when a salt is formed, we can write the equation in these three ways:

1 Word equation

hydrochloric acid + sodium hydroxide →

sodium chloride + water

2 Balanced equation

$HCl(aq) + NaOH(aq) \rightarrow NaCl(aq) + H_2O(l)$

3 Ionic equation

$H^+(aq) + Cl^-(aq) + Na^+(aq) + OH^-(aq) \rightarrow$

$Na^+(aq) + Cl^-(aq) + H_2O(l)$

If we cross out the ions that are the same on both sides of this equation, it becomes:

$H^+(aq) + OH^-(aq) \rightarrow H_2O(l)$

All reactions where an acid reacts with an alkali to form a salt and water can be written like this. In this case only two ions were actually involved in the reaction $H^+(aq)$ and $OH^-(aq)$.

D. Interpret

- Write ionic equations for the reactions between the following substances. Simplify the equations as much as possible.

  a nitric acid and sodium hydroxide;

  b sulphuric acid and potassium hydroxide;

  c silver nitrate and sodium chloride to form a precipitate of silver chloride;

  d copper(II) chloride and sodium hydroxide to form a precipitate of copper(II) hydroxide.

Extension
More about oxidation and reduction

So far we have used the word oxidation to describe a reaction in which oxygen is added to a substance during a reaction. Oxidation can also be used to describe reactions in which hydrogen is removed from a substance.

The word reduction is used to describe a reaction in which hydrogen is added to a substance or oxygen is removed from a substance during a reaction.

An electron micrograph of the rusty bodywork of a car. The top (blue) layer is respray paint bonded to the original paint (green) of the car. The rest is rust! Rusting is an oxidation process

The reaction involved in using a breathalyser is a reduction process

Let us look closely at the reaction in which magnesium is burnt in air to form magnesium oxide. The magnesium is oxidised as it gains oxygen.

magnesium(s) + oxygen(g) →

magnesium oxide(s)

We can write an ionic equation to show what happens to the electrons in this reaction:

$Mg(s) + \frac{1}{2}O_2(g) \rightarrow Mg^{2+}O^{2-}(s)$

This can be written more simply as:

$Mg - 2e^- \rightarrow Mg^{2+}$

$\frac{1}{2}O_2 + 2e^- \rightarrow O^{2-}$

The magnesium atom has lost two electrons and these have been given to the oxygen atom to form magnesium and oxide ions. The magnesium atom has been oxidised. The oxygen atom has gained two electrons: it has been reduced. Oxidation is the chemical process of losing electrons. Reduction is the chemical process of gaining them. These two words will help you to remember this: 'oil rig' (**oxidation is loss, reduction is gain**).

You can now see why oxidation and reduction reactions are usually grouped together. If one occurs, then so must the other. When you tried to construct your reaction tree you may have had difficulty deciding which reactions were oxidation and which were reduction. This is because many of the reactions could be called oxidation *or* reduction reactions, depending on which reactant you were considering. In fact, these reactions are sometimes called redox reactions (**red**uction-**ox**idation reactions).

Oxidation can be

a the gain of oxygen;

b the loss of hydrogen;

c the loss of electrons.

Reduction can be

a the gain of hydrogen;

b the loss of oxygen;

c the gain of electrons.

E. Think about **EXTENSION**

- For each of the following reactions write:

  a a word equation;

  b a symbolic equation;

  c an ionic equation;

  d ionic equations to show electron loss and electron gain.

For each reaction say which reactant is being oxidised and which is being reduced.

1 Magnesium burning in carbon dioxide to form magnesium oxide and carbon.

2 Carbon reacting with copper(II) oxide to form copper and carbon dioxide.

3 Magnesium being added to hydrochloric acid to form magnesium chloride and hydrogen.

4 Carbon monoxide being added to iron(III) oxide in the blast furnace to form carbon dioxide and iron.

5 Zinc being added to a solution of silver nitrate to form a solution of zinc nitrate and a precipitate of silver.

ELECTROCHEMISTRY

Pure water hardly conducts electricity at all. However, if a small amount of an ionic substance is dissolved in the water, it will conduct electricity. Water can then be decomposed into its elements of hydrogen and oxygen by passing electricity through it. Tap water contains ionic impurities, so you can do this experiment yourself at home. The photograph shows you how.

⚠️ You *must* use a battery and *not* the mains electricity.

Decomposing water into its elements. Can you explain what is happening?

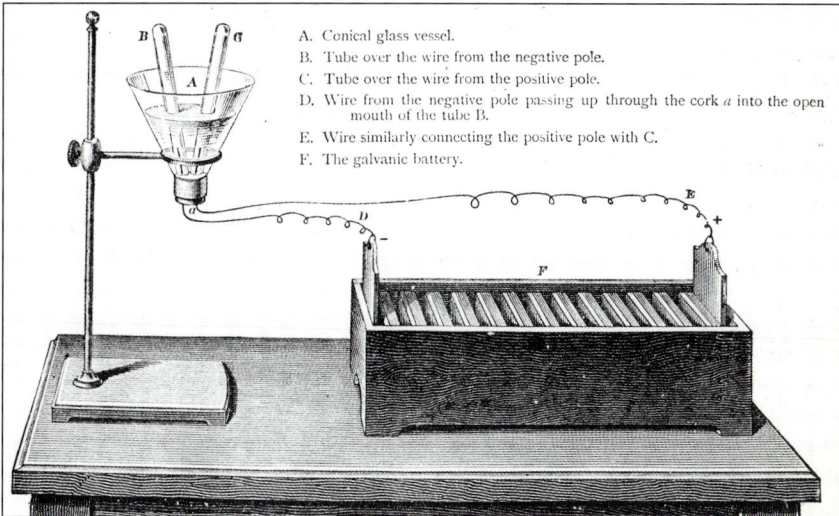

A. Conical glass vessel.
B. Tube over the wire from the negative pole.
C. Tube over the wire from the positive pole.
D. Wire from the negative pole passing up through the cork *a* into the open mouth of the tube B.
E. Wire similarly connecting the positive pole with C.
F. The galvanic battery.

Apparatus for repeating Carlisle and Nicholson's experiment in which they decomposed water into its elements using a voltaic pile

The electrolysis of water was the first important electrolysis experiment to be carried out. It was first done in 1801 by William Nicholson and Anthony Carlisle.

In *Chemical processes* Unit 4 you saw how important the electrolysis of salt is and learnt some rules for predicting what happens at each electrode during electrolysis. Do you remember them? (If not look back!)

Lead bromide is a white powder which does not conduct electricity. However, when it is heated it melts and then the liquid lead bromide does conduct electricity. You should be able to predict why this happens. Annisa watched her teacher demonstrate this experiment in a fume cupboard. She decided to describe what happened from the point of view of an electron called Ernie. Her account is shown on page 82.

The electrolysis of lead bromide by Ernie the electron

I'm Ernie, I live in the outside shell of a lead atom. There are two of us together in this outside shell, me and my friend Cuthbert. We rattle around in a huge space.

Recently, the atom got really hot and started moving very quickly. I wondered what on earth was going on.

Suddenly, a bromine atom rushed up towards us and one of the seven cheerful electrons called out 'Come and join us. We've got room for another in here.'

I jumped into their shell. My friend Cuthbert jumped into the outside shell of another bromine atom. It all went quiet then, but my home seemed quite different.

After a few days it got really hot again, all the electrons in my new home were moving faster and faster all the time. I was moving towards a huge black rod, then I seemed to be dragged right up it. It was really dark. I seemed to be in a really long tunnel. The rest of the bromine atom where I had been living was left behind.

Finally, I came out at another different black rod. I saw lots of lead particles and believe it or not I saw my old home. I quickly jumped down into the outside shell. I don't know where Cuthbert went, I had a new friend. She said her name was Esmerelda! That's how the two of us came together.

F. Think about

- Read the account of Ernie carefully and then answer these questions:

  1 Ernie was an electron. Was he a positive or negative particle?

  2 What element did Ernie belong to, at first?

  3 What took Ernie away from his home?

  4 What was the name of the 'new' solid that Ernie then found himself in?

  5 When Ernie joined his new home, what new particle was made?

  6 What were the tunnels that Ernie went through?

  7 A gas was seen on the electrode that Ernie went to, what was the name of the gas?

  8 A metal is coating the other electrode. What was the metal?

  9 What are the symbols for:

     a lead;

     b bromine?

  10 What is the formula of lead bromide?

Unfortunately it is not easy to carry out the electrolysis of molten solids in the lab. It is not possible to get them hot enough to melt them. Fortunately many ionic compounds are soluble and aqueous solutions of them conduct electricity easily. However, when we have an aqueous solution we have also to consider the two elements hydrogen and oxygen.

G. Work out

- You are given a selection of ionic solutions. Your task is to predict what will happen when an electric current is passed through each of them, and then to check if your predictions are correct.

- Draw up a table like this one before you begin:

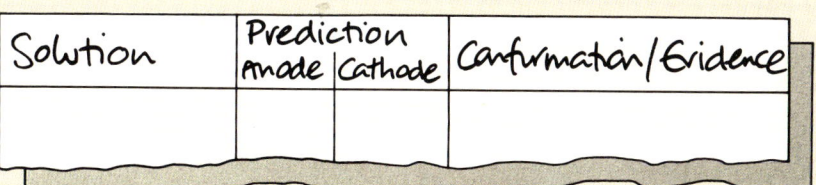

Solution	Prediction		Confirmation/Evidence
	Anode	Cathode	

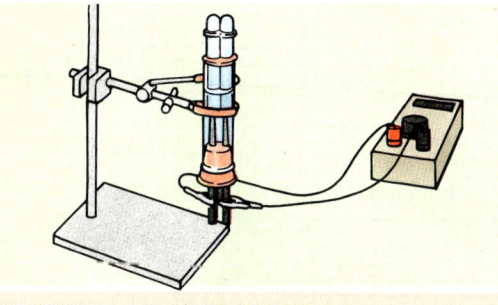

- You should arrange your apparatus like this:

  Your teacher may suggest an alternative arrangement for the electrolysis cell.

- Write a report of your investigation in which the following points are included:

  a   What your predictions were in each case.

  b   How you confirmed these predictions. You will need to include your evidence. This might be from your experiment or, in some cases, you might have to look up data. Remember to quote your sources if you use second-hand evidence.

  c   Why you choose the quantities you did for your experiment (for voltage etc.).

  d   The limitations of your experiment.

  How accurate was it?

  What would you do to improve it?

  e   Suggestions for other similar solutions that you would like to investigate and your predictions as to what might form at each electrode when they are electrolysed.

When you use copper(II) sulphate as an electrolyte, the copper is deposited at the cathode. This is called electroplating. One of the uses of electroplating is that it is possible to coat a cheaper metal with a thin layer of a more expensive one. You may own some gold or silver plated jewellery, which is made this way.

A rack of copper-plated sheets being removed from an electroplating bath during the first stage in the preparation of flexible electronic circuits

A company involved in electroplating has a number of expenses to consider when deciding if a process is viable. Wolfson, a scrap metal company know that they can obtain pure copper from impure copper by electroplating. Your group has been given the task of working out the efficiency of the process involved. The company's financial division will put in the cost but you must investigate the factors that could make copper cheaper.

H. Investigate

- Discuss the factors that may influence the amount of copper deposited at the cathode.

- When you have a list of factors discuss them with your teacher.

- Investigate each of the factors. You must make sure that you control other variables that may affect the result. Keep detailed records during the investigation. You will need to submit these with the report.

- Write a report to the company suggesting:

  a   the factors which increase the amount of copper appearing at the cathode;

  b   other factors that will need to be considered before the process can be considered economical.

- Include your clear records and observations as an appendix. Remember that the directors may not be scientists, but should be able to follow what you did.

Extension

It is possible to increase the amount of an element that is formed at an electrode by increasing the quantity of electricity flowing. The quantity of electricity is measured in coulombs:

1 amp × 1 second = 1 coulomb

I. Work out

- Calculate how much electricity is needed to deposit 1 mole (64 g) of copper atoms by carrying out the experiment on Worksheet PB9.

- In similar experiments the following results have been obtained:

Element	Number of coulombs needed to form one mole of atoms at an electrode
aluminium	289 500
hydrogen	96 500
lead	193 000
silver	96 500
zinc	193 000

Plot a bar chart which shows these results and your own results for copper.

What pattern can you see from the bar chart?

Accurate experiments have shown that 96 500 coulombs of electricity are needed to deposit one mole of silver. In fact, any ion with a single positive charge, such as Ag+ needs 96 500 coulombs of electricity to liberate one mole of atoms at the cathode. One mole of electrons is called a Faraday, it is equivalent to 96 500 coulombs. Looking back at the table, it can be seen that aluminium needs three moles of electrons. Put simply, the aluminium ion is Al^{3+}.

J. Think about

- Copy the following table into your book:

Element	Number of moles needed to form 1 mole of atoms at the electrode	
aluminium	3	
copper		
hydrogen		
lead		
silver	1	1+
zinc		

- Complete the middle column.
- For each element write in the end column the charge on one of its ions.

  (Think carefully so that you get the sign correct.)
- Add a title to this end column.

Electrolysis experiments can give important information about the quantities involved. Three factors are important:

a   the amount of electricity used;

b   the amount of element deposited;

c   the charge on the ions.

As long as two of these factors are known the other can be calculated.

Electroplating ornate pots

UNIT 4
THE CHANGING NUCLEUS

Section 4.1

DISCOVERING RADIOACTIVITY

Radioactivity was first discovered by a Frenchman, Henri Becquerel, in 1896. Marie Curie did much of the early investigatory work and she was the first to use the term radioactivity in 1898.

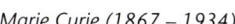

Marie Curie (1867 – 1934)

Earlier units discussed a model of the nucleus of an atom as a collection of protons and neutrons. We know that electrons take part in chemical reactions and we have assumed that the inner nucleus remains stable. In 1896 Becquerel noticed that the element uranium affected a photographic plate when all he had done was to leave them in the dark together. He wrote:

... I showed that uranium salts emit radiations whose existence had not been recognised and that these radiations showed some remarkable properties... The radiations of the uranium salts are emitted not only when the substances are exposed to light, but even when they are kept in darkness, and for more than two months the same fragments of various salts, shielded from all the exciting radiations known, have continued to emit the new rays, almost without perceptible weakening...

He concluded that the rays were given out by the uranium. It is these rays that Marie Curie later called radioactivity. Since then it has been found that atoms of many elements have unstable nuclei and can emit particles or energy either spontaneously or when bombarded by other atoms.

A. Discuss

1 What do you think 'radioactivity' is?

2 What is the link of each of the objects in the picture with radioactivity?

3 Discuss whether each of these statements about radioactivity is true or false:

  a We are continually exposed to background radiation.

  b About 40% of the total background radiation is made by humans.

  c The Sievert is the unit of radiation dose.

  d Cosmic rays from space contribute to our radiation dose.

  e Radiation can increase the chances of a heart attack.

  f A radiation worker is more likely to die of cancer than a deep sea fisherman is to have an accident during their working life.

  g The National Radiological Protection Board sets limits for doses of radioactivity.

  h There is only one type of radioactivity.

  i Elements can change into new elements through radioactive decay.

  j All elements with an atomic mass greater than bismuth are radioactive.

The alchemists dreamed of turning 'base' metals into gold. We now know that certain elements can be changed into others through radioactive decay. This is called transmutation.

We are going to consider three types of ray given out by radioactive material.

Alpha (or α) particles

An alpha particle is made up of two protons and two neutrons. An alpha particle is actually a helium nucleus. The particle is positively charged. The parent atom which gives off an alpha particle changes into a new type of atom. For example, radium is changed into radon:

$$^{226}_{88}Ra \rightarrow ^{2}_{86}\text{к}n + ^{4}_{2}He \text{ (alpha particle)}$$

Alpha particles only travel a few centimetres from the parent atom and can be stopped by a thin piece of paper. However, they are quite big particles and as they are positively charged they attract electrons from other atoms. They are highly ionising.

Beta (or β) particles

Beta particles are electrons. The radiation is formed as a nucleus of an atom gives off an electron. This happens when a neutron in the nucleus turns into a proton and an electron. The electron is ejected from the nucleus:

$$n \rightarrow p^+ + e^- \text{ (beta particle)}$$

This beta particle has a negative charge and travels at almost the speed of light. The parent atom changes into an atom of a different element as it loses this beta particle. For example strontium changes into ytterbium. The mass number of 90 does not change, because the proton which remains in the nucleus has the same mass as the former neutron:

$$^{90}_{38}Sr \rightarrow ^{90}_{39}Y + e^- \text{ (beta particle)}$$

Beta particles do not cause as much ionisation as the alpha particles but they do travel further. They are stopped by a thin sheet of aluminium.

Gamma (or γ) rays

γ-rays are electromagnetic radiations like X-rays and light. These are given out from a nucleus when it loses an alpha or a beta particle as well. They are high energy rays and cause less ionisation than either alpha or beta particles. However, due to their short wavelength and high energy they need several metres of concrete to stop them.

Measuring radioactivity with a Geiger counter and scaler

B. Work out

• Copy and complete this chart summarising the three types of radiation:

Name	How formed from an atom	Equation	How penetrating	How ionising
alpha				
beta				
gamma				

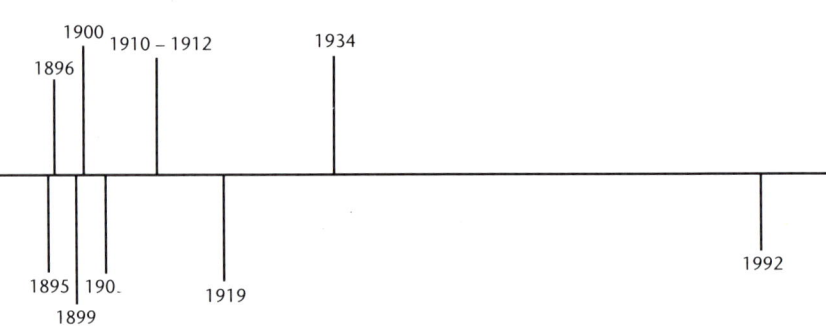

This time line shows some important dates in the discovery of radioactivity

C. Research

• Draw your own time line showing the important dates in the discovery of radioactivity.

• Find out about the contributions that the following people made to radioactivity and add their names and brief details of their work to your time line:

  Becquerel   Marie and Pierre Curie   Röntgen   Rutherford   Soddy   Villard

  (Keep a note of your references, you may need more details about them later in the unit.)

• Collect at least five examples of the uses of radioactivity. You will need them later in the unit.

PATTERNS OF DECAY

Tim's Mum had been feeling really tired and lazy for the past few months. She could not be bothered to do anything, and she seemed so forgetful. She never used to be like that. Tim remembered when she was always busy and went out a lot. Eventually Tim's mum was persuaded to visit her GP. The GP suspected thyroid inactivity or myxodoema. He noticed that her skin was dry and she confirmed that she had been losing quite a bit of hair recently. These are symptoms of thyroid inactivity. After some months Mrs Coleman had an appointment with a consultant at the hospital. The consultant explained to Mrs Coleman that it was necessary to check on her thyroid activity by using a radioactive tracer. She would take this tracer orally and this would enable the thyroid to be monitored. Tim's mum looked worried but the consultant explained there was nothing to worry about since the isotope used as the tracer had a half-life of only 13 hours. Mrs Coleman left still feeling rather anxious. It seemed that the test would not hurt, but what was an isotope and what was a half-life?

What are isotopes?

Isotopes are neither rare nor special. Most elements have several isotopes. The word is just used to describe atoms of the same element which have different masses. In Mrs Coleman's case she was to take the isotope iodine-123. This can be written as:

$$^{123}_{53}I$$

Remember this is the mass number, which tells you how many protons and neutrons there are

Remember this is the atomic number, which tells you how many protons (or electrons) there are

Other isotopes of iodine include iodine-131, which is also used in thyroid diagnosis because it is radioactive, and iodine-127, which is the commonest sort used in schools.

What is a half-life?

The nuclei of the iodine-123 that Mrs Coleman has to take are unstable. As the particles in the nucleus rearrange themselves gamma rays are emitted which will be detected. It is not possible to predict precisely when any one particular nucleus will change, but it is possible to measure which fraction of the atoms will break up (or decay) in a given time. The time required for half of the atoms of a radioactive isotope to break up is called the half-life. For the iodine-123 that Mrs Coleman has to take the half-life is 13 hours. This can be shown as:

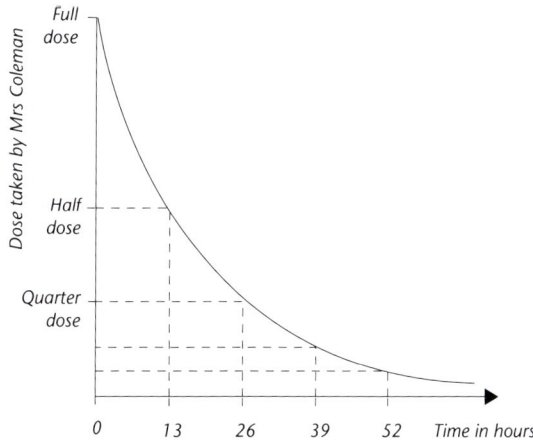

The half-life of iodine-123 is 13 hours

Each isotope has its own half-life. Some isotopes decay more quickly than others. Some take thousands of years. The half-life of C^{14} is about 6000 years.

The diagram shows the apparatus that can be used to measure the half-life of an isotope.

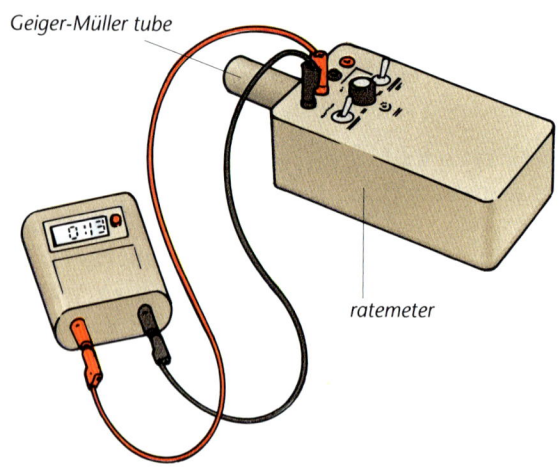

Geiger-Müller tube

ratemeter

These are some results that were obtained when an experiment was carried out to find the half-life of protactinium-234.

Half-lives can vary from millionths of a second to millions of years. Some half-lives are shown in this table.

Isotope	Half-life	Type of decay
carbon-14	6000 years	β
cobalt-60	5 years	γ
strontium-90	28 years	β
plutonium-239	24 000 years	α, γ
uranium-238	5×10^9 years	α
iodine-123	13 hours	α
lawrencium-140	14 seconds	β
strontium-93	8 minutes	β, γ
strontium-94	14 minutes	β, γ
radon-222	4 days	α

Knowing both the half-life of an isotope and the type of radiation it emits is important in choosing isotopes for different tasks.

Time (s)	Counts per minute
0	–
60	2360
90	1880
120	1440
150	960
180	800
210	missed reading
240	480
270	400
300	240

F. Interpret

- Imagine that you are a journalist. You are visiting a local company which has a licence to irradiate herbs and spices. The design team has drawn this diagram for you:

Inside a food irradiation plant

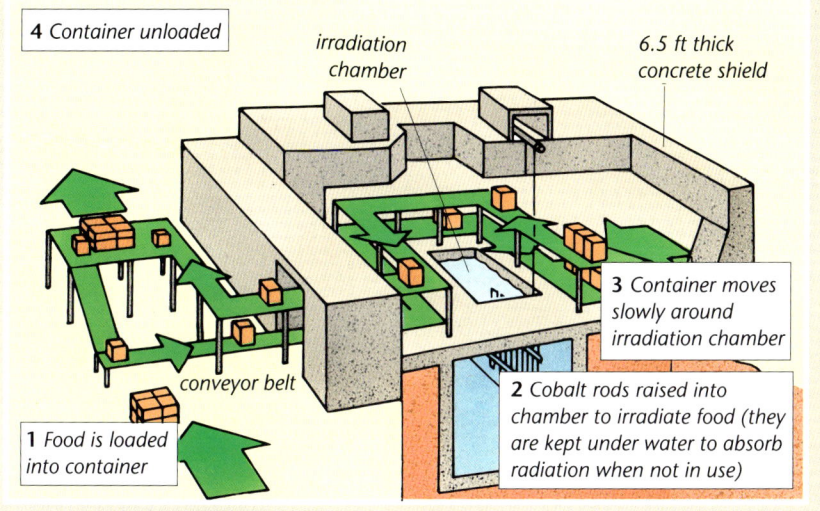

4 *Container unloaded*

irradiation chamber

6.5 ft thick concrete shield

3 *Container moves slowly around irradiation chamber*

conveyor belt

2 *Cobalt rods raised into chamber to irradiate food (they are kept under water to absorb radiation when not in use)*

1 *Food is loaded into container*

Write a suitable text to accompany the diagram for submission to the editor. In your text you must:

a   explain how the process works;

b   allay local fears about the process.

- Add a note suggesting a suitable title for the article and another note explaining why you think the paper should run the story.

G. Research

- Copy and complete the following table to suggest both the nature of the radiation and the isotope that could be used for the various tasks. You may not be able to name all the the isotopes but you should be able to complete the 'reasons for choice' column and indicate the range of half-life that would be accceptable.

Task	Radiation (α β γ)	Possible isotope	Reason for choice
preserving foodstuffs			
sterilising syringes			
detecting aircraft cracks			
studying kidney function			
following river effluent flow			
studying brain function			

H. Work out

1   The half life of plutonium-239 is 24 000 years. What fraction of the example will remain after 24 000 years?

2   Strontium-90 may be absorbed into human bone. Its half-life is 28 years. How much of the original amount will be left after 56 years?

3   An isotope carbon-14 has a half-life of about 6000 years. The amount of this isotope in a living organism decreases from the date of death because no fresh carbon-14 is incorporated into the tissues once it has died. This is the principle behind the radiocarbon dating of archeological remains. What fraction of the original amount of carbon-14 will remain after 24 000 years?

4   The half-life of iodine-131 is eight days. Of an original 12 g sample only 1.5 g remained. How long had the sample been kept?

5   Soot from the outside of a cooking pot was radiocarbon dated. It was found that compared to living material only 25% of the carbon-14 remained.

a   How old was the soot?

b   Why is this not an accurate method for dating the cooking pot?

NUCLEAR ENERGY

The nucleus of an atom contains protons and neutrons. Large nuclei with a proton number of over 50 tend to be more unstable than those with fewer protons. Large numbers of protons tend to repel each other and this causes the nucleus to break apart into two large fragments. This process is called nuclear fission. When uranium-235 undergoes nuclear fission it splits roughly in half and two or three neutrons are shot out and a tremendous amount of energy is released. Nearby Uranium-235 atoms may be hit by the escaping neutrons, and this can cause more nuclear fission. More neutrons and energy are produced, causing more nuclei to decay, and an explosive chain reaction can occur.

Explosion of an atom bomb in the South Pacific, in October 1952

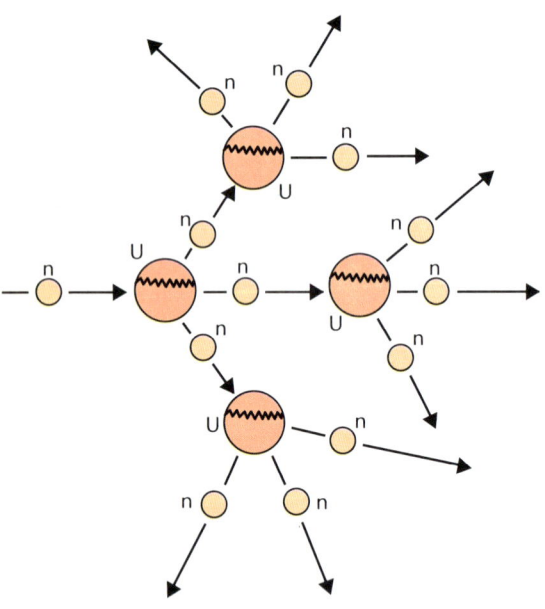

An idealised chain reaction

A nuclear fission reaction is very fast. It occurs in a fraction of a second. The explosive charge of an atomic bomb contains almost 100% uranium-235. This is held in a strong container. Detonation causes the uranium-235 to be compressed. The chain reaction begins. The neutrons cannot escape without hitting other uranium-235 atoms and so millions of uranium-235 atoms undergo nuclear fission. An explosion results from the huge amount of energy that builds up so quickly.

I.   Think about

- A bad explosion occurred at a local waste disposal site and a man was blown from the top of a lorry that he was welding. In the report in the local paper, the account included the phrase *'It was just like an atomic bomb'*. You know that this is just a journalist's phrase but you feel strongly that when science is mentioned in the newspaper the facts should be correct. Prepare a short account that the paper could print explaining how an atomic bomb explosion is caused.

- Try to explain how the explosion with the lorry could have been caused.

On the 6th of August 1945 the first atomic bomb to be used on people was dropped on Hiroshima in Japan. It was equivalent to 12.5 thousand tons of high explosive and 130 000 people had died as a result of this bomb by the end of October 1945. This and the following bomb on Nagasaki three days later resulted in the surrender of Japan on 14 August 1945.

A nuclear reactor

Scientists involved in nuclear fission work during the Second World War realised that the heat produced during the fission period could be used to generate electricity. In the decay of uranium-235 two or three neutrons are formed as the uranium nucleus disintegrates. If only one of these is allowed to bombard other uranium nuclei then the chain reaction can be slowed down and controlled and the amount of energy produced can also be controlled. To do this it is necessary to absorb the other neutrons. This is exactly what happens in a nuclear reactor. The uranium fuel is divided up into portions with a neutron absorbing substance, such as boron, placed in between the portions to absorb the excess neutrons.

There are many kinds of nuclear reactor in the world but they all work on the same principle. The heat needed to turn water into steam, and so drive the turbine which generates the electricity, comes from the energy released when the nucleus of a heavy atom splits up.

In coal- and oil-fired power stations this heat is produced by a chemical reaction when the fuel burns.

J. Design

• Imagine that you are going to visit a nuclear power station. Design some material that it would be useful for you to have before your visit. The material must include an explanation of what nuclear fission is and how it can be used to produce electricity. The material could be a leaflet, a worksheet, self-learning text, clear diagrams, a model, a set of slides or a video. If you choose one of the last two options you will need to provide detailed plans or a storyboard to show what you would produce.

Today, nuclear power accounts for nearly 20% of the energy that is generated in the UK. Uranium, strictly speaking, is a non-renewable source of energy.

- natural gas 2.5%
- uranium 4%
- oil 7%
- coal and peat 86.5%

The approximate known fuel reserves

The world demand for energy is increasing. The population is increasing rapidly and we are using more electrically-operated devices as our standard of living increases. Providing this energy cheaply without causing environmental damage is difficult. There is no single correct solution to this problem but a number of options are available.

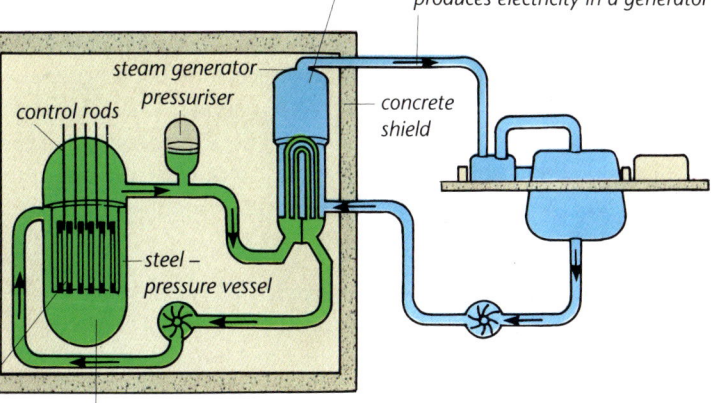

The control rods are made of boron and can absorb neutrons. Moving them up will increase the energy coming out of the reactor – pushing them in reduces it. The control rods can absorb all the neutrons and shut down the reaction completely

*Water is heated by the hot water from the reactor core. The two sorts of water do **not** mix. Steam is generated*

Steam travels to a turbine which produces electricity in a generator

steam generator
pressuriser
concrete shield
control rods
steel – pressure vessel

Light water in the reactor core is at a high pressure. This is heated up by the nuclear fission reactions

Fuel elements (uranium dioxide clad in alloy of zirconium). The fuel only contains 3.2% uranium-235

***Notice** how the water from the reactor core and steam generator are recycled*

A pressurised water reactor built at Sizewell B in Suffolk

K. Discuss

1 How can individuals reduce their energy needs?

2 How can communities reduce their energy need?

3 What alternative sources of energy can be developed to take the place of that supplied by fossil fuels at present?

4 Why is it so difficult to convince people of the energy problems which we will face in the future?

RADIOACTIVE ALARMS

Statistics show that in fires many more people are killed by the smoke than by the flames. This is quite a problem to many large shops where large amounts of dense toxic fumes could build up quickly if a fire started. One large chain of department stores has just begun a programme of installing smoke alarms in its shops.

The inside of a smoke detector

Firemen fighting a blaze in East London. More people are killed in fires by the smoke than by the flames

Ionisation smoke alarms contain a small amount of americium-241. This radioactive isotope gives off alpha particles which ionise the air inside the alarm. These ions allow a small current to flow from the battery. Should smoke enter the detector the current flow is reduced and the alarm is triggered.

Many of the staff are anxious about the prospect of working near smoke alarms. They have heard that the smoke alarms are radioactive.

Annual dose of radiation

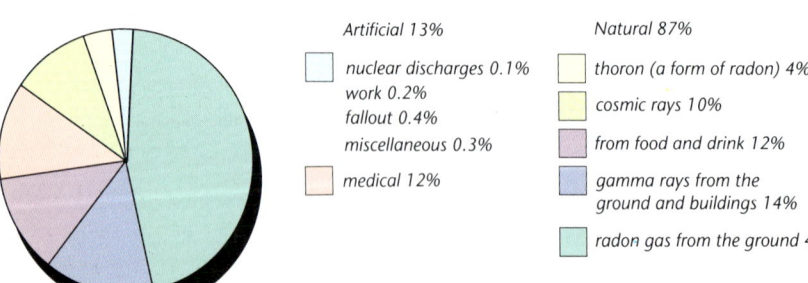

Artificial 13%

☐ nuclear discharges 0.1%
work 0.2%
fallout 0.4%
miscellaneous 0.3%

☐ medical 12%

Natural 87%

☐ thoron (a form of radon) 4%

☐ cosmic rays 10%

☐ from food and drink 12%

☐ gamma rays from the ground and buildings 14%

☐ radon gas from the ground 47%

The average annual dose to the UK population is 2.5 millisieverts overall, but there are great variations from the average

L. Work out

- Imagine that your team has been invited to plan a presentation to inform the staff of a large store about radioactivity. Consider your audience:

  What is likely to be the most effective way of presenting the information? Is it likely to be a seminar, a leaflet for all the staff, a workshop (where the staff would do something) or will a variety of methods be needed?

  What will you be trying to tell them about?

  What background information will the staff need?

- Plan your presentation. If necessary produce leaflets, posters etc. Your teacher will tell you how much detail is needed. You can use information that you have met in this unit but, where possible, supplement it with data from other sources.

- If it is possible, try out your ideas on others in the class, or better still on people who have not attended the lessons about radioactivity. (Discuss this with your teacher.)

- Write an evaluation sheet that the 'audience' can fill in.

- Write a brief report of your ideas including points made in the evaluation and how you could adopt the presentation for future use.

UNIT 5
THE PERIODIC TABLE

Section 5.1

GETTING THE MOST OUT OF THE PERIODIC TABLE

The periodic table has been referred to several times in this book, and you have used it to obtain some information about substances. What exactly is the periodic table and what other information can you get from it?

The periodic table contains 92 naturally occurring elements as well as the artificial ones that have been made since 1940, when the first artificial element, Neptunium, was made.

One version of the periodic table is given here. You will find out more about how to use it in this unit.

The periodic table

Group	I Alkali metals	II Alkaline earth metals											III	IV	V	VI	VII Halogens	0 Noble gases
Period 1						1 H Hydrogen 1.0												2 He Helium 4.0
2	3 Li Lithium 6.9	4 Be Beryllium 9.0											5 B Boron 10.8	6 C Carbon 12.0	7 N Nitrogen 14.0	8 O Oxygen 16.0	9 F Fluorine 19.0	10 Ne Neon 20.2
3	11 Na Sodium 23.0	12 Mg Magnesium 24.3											13 Al Aluminium 27.0	14 Si Silicon 28.1	15 P Phosphorus 31.0	16 S Sulphur 32.1	17 Cl Chlorine 35.5	18 Ar Argon 39.9
4	19 K Potassium 39.1	20 Ca Calcium 40.1	21 Sc Scandium 45.0	22 Ti Titanium 47.9	23 V Vanadium 50.9	24 Cr Chromium 52.0	25 Mn Manganese 54.9	26 Fe Iron 55.9	27 Co Cobalt 58.9	28 Ni Nickel 58.7	29 Cu Copper 63.5	30 Zn Zinc 65.4	31 Ga Gallium 69.7	32 Ge Germanium 72.6	33 As Arsenic 74.9	34 Se Selenium 79.0	35 Br Bromine 79.9	36 Kr Krypton 83.8
5	37 Rb Rubidium 85.5	38 Sr Strontium 87.6	39 Y Yttrium 88.9	40 Zr Zirconium 91.2	41 Nb Niobium 92.9	42 Mo Molybdenum 95.9	43 Tc Technetium (99)	44 Ru Ruthenium 101.1	45 Rh Rhodium 102.9	46 Pd Palladium 106.4	47 Ag Silver 107.9	48 Cd Cadmium 112.4	49 In Indium 114.8	50 Sn Tin 118.7	51 Sb Antimony 121.8	52 Te Tellurium 127.6	53 I Iodine 126.9	54 Xe Xenon 131.3
6	55 Cs Caesium 132.9	56 Ba Barium 137.3	57 La Lanthanum 138.9 ▶	72 Hf Hafnium 178.5	73 Ta Tantalum 181.0	74 W Tungsten 183.9	75 Re Rhenium 186.2	76 Os Osmium 190.2	77 Ir Iridium 192.2	78 Pt Platinum 195.1	79 Au Gold 197.0	80 Hg Mercury 200.6	81 Tl Thallium 204.4	82 Pb Lead 207.2	83 Bi Bismuth 209.0	84 Po Polonium (210)	85 At Astatine (210)	86 Rn Radon (222)
7	87 Fr Francium (223)	88 Ra Radium (226)	87 Ac Actinium (227) ▶▶	104 Unq Unnilquadium (261)	105 Unp Unnilpentium (262)	106 Unh Unnilhexium (263)												

▶ Lanathanoid elements

58 Ce Cerium 140.2	59 Pr Praseodymium 140.9	60 Nd Neodymimum 144.2	61 Pm Promethium (147)	62 Sm Samarium 150.4	63 Eu Europium 152.0	64 Gd Gadolinium 157.3	65 Tb Terbium 158.9	66 Dy Dysprosium 162.5	67 Ho Holmium 164.9	68 Er Erbium 167.3	69 Tm Thulium 168.9	70 Yb Ytterbium 173.0	71 Lu Lutetium 175.0

▶▶ Actinoid elements

90 Th Thorium 232.0	91 Pa Protactinium (231)	92 U Uranium 238.1	93 Np Neptunium (237)	94 Pu Plutonium (242)	95 Am Amercium (243)	96 Cm Curium (247)	97 Bk Berkelium (245)	98 Cf Californium (251)	99 Es Einsteinium (254)	100 Fm Fermium (253)	101 Md Mendelevium (256)	102 No Nobelium (254)	103 Lr Lawrencium (257)

Key

Atomic number
Symbol
Name
Relative atomic mass

- reactive metals
- metalloids
- non-metals
- noble gases
- transition metals (most of the metals we use are found here)

A. Think about

- Decide which of the following information you can find out from the periodic table. Give answers where you can.

1 The symbol for the element silicon

2 The atomic mass of chlorine

3 The number of protons in argon

4 The name of the element with the symbol Mn

5 The number of neutrons in the nucleus of potassium

6 How many electrons there are in the outer shell of oxygen

7 The name of an element that is similar in many ways to sodium

8 The name of a noble gas present in the air

9 The formula of the compound formed when sodium burns in chlorine

10 The name of an element that is likely to form compounds containing covalent bonding.

There are many similarities between using the periodic table and buying a CD. Think how the music counter is usually arranged in a shop. LP records, CDs and tapes are usually given separate sections. In the CD section the CDs are usually divided up into further sections such as pop, jazz, film theme tunes, heavy metal and folk. In each of these sections you will probably find that the CDs are arranged alphabetically. For example, you would find a particular Beatles CD under the letter 'B' in the pop section.

In what ways is buying a CD similar to using the periodic table? In the periodic table elements which are similar to one another are grouped together. By knowing some features about the element you are able to locate it in the periodic table. The features of one element will also tell you something about nearby elements. Similarly at the music counter, by knowing the characteristics of the 'Beatles' type of music you are able to locate their CD. This CD will be found in the larger group of 'pop' CDs which contains broadly similar types of music.

Any model that is used always has its limitations. If you know quite a bit about the periodic table already you will be able to spot some of the problems of using the music shop model! If not, you may like to see if you can spot where the music shop model goes wrong when you have finished the unit!

By knowing how the music shop has arranged its stock you are able to predict where to look for a CD. It is the same with the periodic table. If you know how the elements within the periodic table are arranged then you will be able to make some predictions about them and their compounds.

Getting started with the periodic table

The previous activity showed you some of the information that you can get from the periodic table. By learning a few basic rules you will be able to use it to give you much more information. These basic rules are as follows:

1. The elements are listed in order of their atomic number.

2. The list has been rearranged into rows. After each noble gas a new row is started.

3. The horizontal rows are called periods. There are seven of these.

4. The vertical columns are called groups. There are eight main groups.

5. The groups have numbers. The numbers tell you how many electrons there are in the outer shell of electrons (for the first 20 elements).

6. The periods also have numbers. These tell you how many shells of electrons the element has.

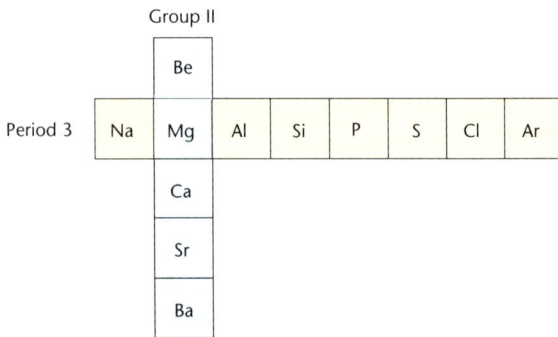

The elements in group II have two electrons in their outer shell. The elements in period 3 have three shells of electrons. Magnesium has two electrons in its third, and outermost, shell

As you move from left to right across period 2 or period 3, you will see that the atomic number, and therefore the number of electrons, increases by one each time. The chemical properties of an element depend on the electrons in its outer shell, so the properties of an element and the way in which it reacts gradually changes as you move across the table.

A group of elements all have the same number of electrons in the outer shell, so all the elements in the group react in a similar way. An atom at the bottom of the group has more electron shells than one at the top so as you move down a group the atoms get bigger.

B. Work out

- Use the information in the table on page 93 to annotate your own copy of the periodic table with the following information:

    a the group numbers and any group names;

    b the period numbers.

- Use different colours to show the elements that are:

    a metals;

    b non-metals;

    c metalloids – sometimes called semi-metals.

- Add your own symbols or notes to show the important trends.

The periodic table can give you a lot of information, if you know the basic rules. If you think about these rules you will see there are many different patterns in the periodic table. For example, the number of electrons increases as you move from left to right across a period, and as you move down a group the atoms get bigger.

In the next activity you will take part in building a class periodic table which will enable you to see the patterns more clearly.

C. Work out

- You are going to make a cube of information about one or more elements of the periodic table. Your teacher will tell you which element(s) to do. First, using the template on Worksheet PB12, cut out a cube. Score along the marked lines.

- Write the following information about the element on the six sides of the cube:

    side 1   name

    side 2   is it a metal, a non-metal or a metalloid

    side 3   symbol

    side 4   state at room temperature (solid, liquid or gas)

    side 5   atomic number, Z

    side 6   relative atomic mass

- Now stick the cube together so that the information is on the outside.

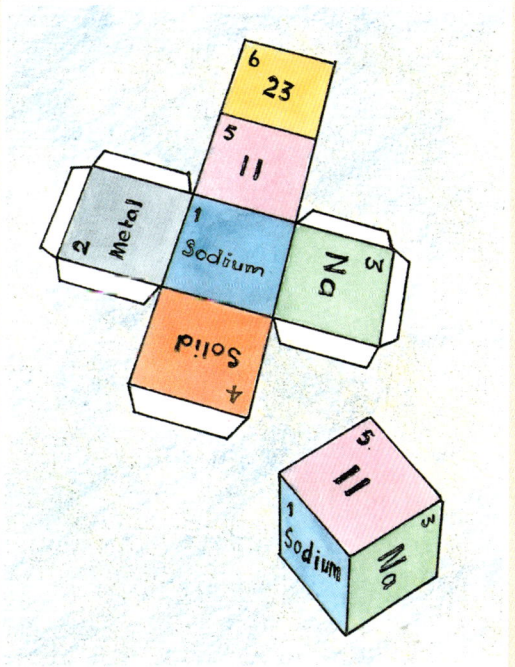

- Place the cube in the correct position on a copy of the periodic table.

- By arranging the cubes on the table with different sides facing upwards you will be able to see the patterns in the periodic table more clearly.

Naming new elements

In the past, when new elements were discovered they were often named after famous scientists. For example, number 96 is Curium and 99 is Einsteinium. However, if you look at elements 104 – 106 you will see some strange names. The discovery of element 104 was claimed by a Russian scientist in 1964. It was called Kurchatovium and given the symbol Ku. When the same element was claimed to be discovered in America in 1969 it was called Rutherfordium and given the symbol Rf. To stop this kind of confusion there is now an agreed system for naming new elements which is based on their atomic number. Element 104 will be called unnilquadium which stands for un = 1, nil = 0 and quad = 4. The symbol Unq is used. Using this system you can understand how unnilpentium (105) and unnilhexium (106) were also named.

USING THE PERIODIC TABLE TO FIND OUT ABOUT GROUPS I AND II

Group I

Group I of the periodic table contains six elements. Sodium is one of these elements. Lithium is another. You met sodium in *Chemical processes* when you studied salt. The elements in group I all contain one electron in their outer shell. It is this outer electron which is important in chemical reactions. By finding out about sodium you can make some predictions about the patterns in behaviour for the other elements in group one.

Lithium and sodium are two elements in group I

D. Interpret

- Use the information in the periodic table to answer the following questions.

1 Give the symbols of two non-metal elements.

2 Name a group containing only metals.

3 Name the element with the largest atom in group II.

4 Name an element with similar reactions to sulphur.

5 Give the symbol of a metalloid.

6 List the elements in period 2.

7 Explain why group 0 is sometimes called group VIII.

8 Say how many electrons there are in the outer shell of boron.

9 Suggest the formula of the compound formed when rubidium reacts with fluorine.

10 Suggest the formula of silicon chloride if carbon can form a chloride with the formula CCl_4.

E. Work out

- Work through these questions about sodium and use your answers to help you to fill in the predictions column for sodium on Worksheet PB13.

1 How is sodium stored? What does this say about its reactivity?

2 How many electrons does sodium have? How are they arranged in shells? What other clue does this give you to its reactivity?

3 Is sodium a metal?

4 What is the charge on a sodium ion?

5 What type of bonding does sodium chloride have?

6 What colour is sodium chloride?

7 What is the formula of sodium chloride?

8 Is sodium chloride soluble?

9 How dense is sodium?

10 What is the pH range of sodium hydroxide solution? The name for group I gives you a clue.

- Confirm your predictions either by watching a demonstration or by using second-hand data.

You cannot base your predictions for all the elements in group I on the information about just one element. You will need some more evidence.

F. Interpret

- Complete the column for lithium on Worksheet PB13.

- Use the results from your completed worksheet to draw up a list of similarities and a list of differences for the elements sodium and lithium.

- Include in the appropriate list the facts that:

  a They react with cold water to form hydrogen and an alkali. For example,
  $$2Na(s) + 2H_2O(l) \rightarrow 2NaOH(aq) + H_2(g)$$

  b They form compounds with similar formulae. For example, NaCl and LiCl.

One of the features of the periodic table is that the atoms get bigger as you go down a group. This means that the outer electrons are further from the positive nucleus and can therefore be removed more easily. Potassium is below sodium in group I but you will not be allowed to do any experiments with potassium metal. Using both these pieces of information you could predict that potassium is more reactive than sodium – and you would be correct. A further prediction about the reactivity of rubidium, caesium and francium would be that as you go down group I the elements get more and more reactive.

In group I, the alkali metals get more reactive as you go down the group.

G. Research EXTENSION

- Find out how the melting points, boiling points and densities of the elements in group I change as you go down the group.

- Describe these patterns.

Group II

The elements in group II are called the alkaline earth metals, which give you two clues as to what they are like.

Some ores which contain elements belonging to group II:

(a) crystals of dolomite (calcium magnesium carbonate) in the foreground with magnesite (magnesium carbonate) behind;

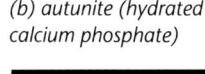

(b) autunite (hydrated calcium phosphate)

Magnesium and calcium are two metals in group II

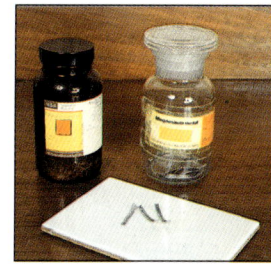

(c) emerald, a green variety of beryl (beryllium aluminium silicate)

H. Work out

- Consider the elements magnesium and calcium in the same way as you did for lithium and sodium in activity E. First make up a table for your predictions and then confirm your predictions by experiment or by using data collected from reference books.

- When you have completed your table, draw up a list of similarities between magnesium and calcium.

I. Think about EXTENSION

- Compare the elements in group I and group II by copying and completing the following table. You should compare either lithium and magnesium or sodium and calcium.

	Group I element	Group II element
How the element is stored		
Reason for this method of storing		
Reaction with water		
Balanced equation for this reaction		
Charge on ion of element		

Are group I elements more or less reactive than the elements in group II?

What evidence do you have for this?

THE HALOGENS – A GROUP OF NON-METALS

The halogens are in group VII of the periodic table. The halogens themselves have some important uses but compounds made from these elements are even more important.

Although the elements in group VII have a number of similarities it is their differences that are the most noticeable when you first see them.

Chlorine, bromine and iodine look very different although they have many similar properties

J. Discuss

- As a group think about the halogens and the information you can get about them from the periodic table.

- Spend about five minutes coming up with the facts and a further five minutes sorting them out into:

  a   things they have in common;

  b   thing they do not have in common.

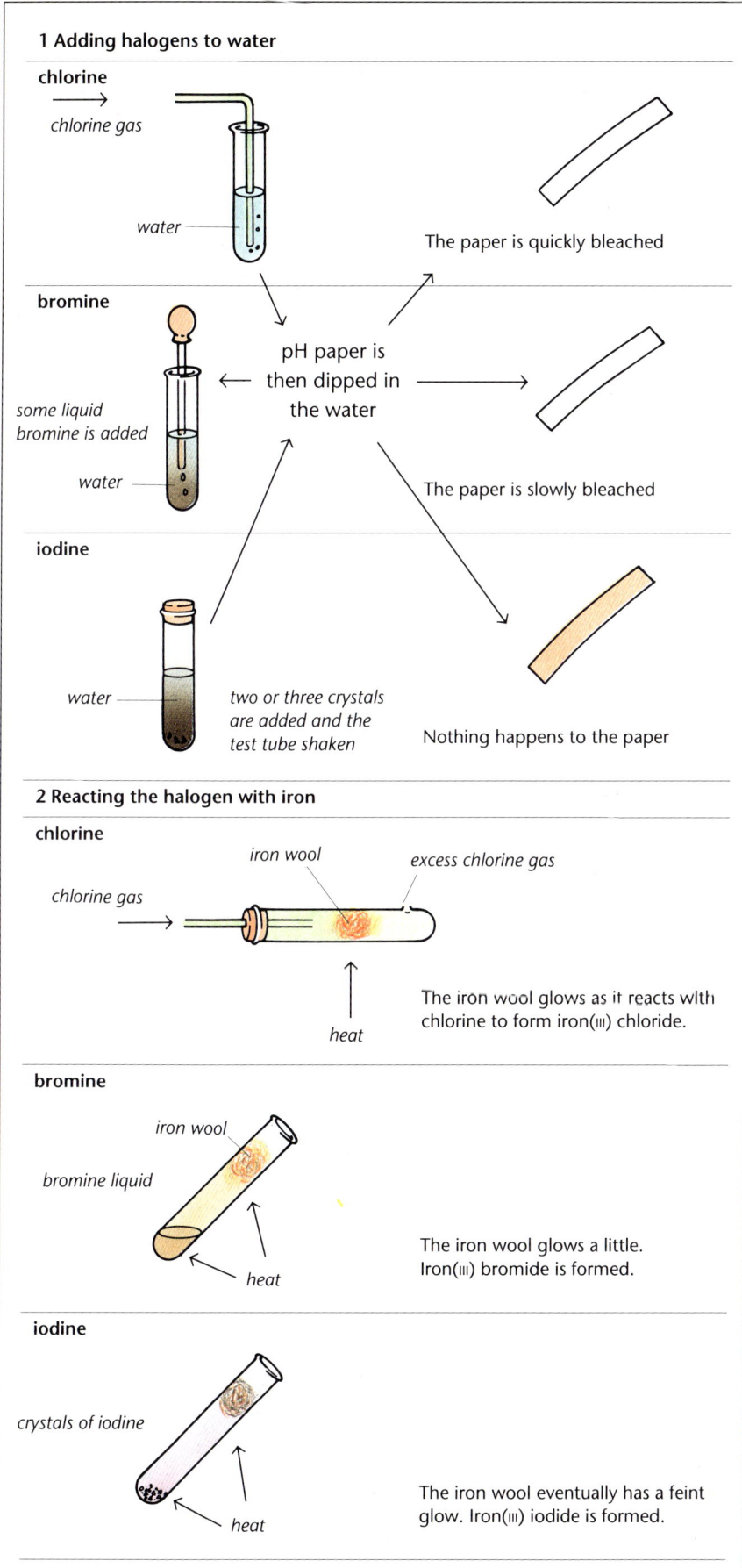

1 Adding halogens to water

chlorine

chlorine gas →

water

bromine

some liquid bromine is added

water

iodine

water

two or three crystals are added and the test tube shaken

pH paper is then dipped in the water

The paper is quickly bleached

The paper is slowly bleached

Nothing happens to the paper

2 Reacting the halogen with iron

chlorine

chlorine gas →

iron wool

excess chlorine gas

heat

The iron wool glows as it reacts with chlorine to form iron(III) chloride.

bromine

iron wool

bromine liquid

heat

The iron wool glows a little. Iron(III) bromide is formed.

iodine

crystals of iodine

heat

The iron wool eventually has a feint glow. Iron(III) iodide is formed.

When you studied the elements in groups I and II you noticed that the reactivity of the elements increased as you went down the group. By investigating a number of reactions of the halogens a trend in this group can be detected too. The diagram shows some details of two experiments with the halogens.

You can also investigate the reactivity of the halogens by carrying out a series of displacement reactions. This is when you add one halogen element to a compound of another halogen and look for a reaction. For example, if we add chlorine gas to a solution of sodium iodide, we can look to see if the chlorine displaces the iodine, to give sodium chloride and iodine. If this did happen, what colour would you expect the solution to become?

If the element chlorine does displace the iodine, this shows that it is more reactive. Displacement reactions have equations of the form:

$$X_2 + 2NaY \rightarrow Y_2 + 2NaX$$

Here element X pushes element Y out so element X is the more reactive.

solution of halogen (bromine, chlorine or iodine) in water

solution of potassium halide (either bromide, chloride or iodide) in water

1, 1, 1 – trichloroethane

Halogen displacement test

In the arrangement shown in the diagram, the colour of the bottom layer, the trichloroethane, indicates which halogen element has been displaced. The halogen elements tend to dissolve more easily in the trichloroethane than the water, so this is where to look for the colour. Bromine gives an orangey colour. Iodine gives a reddish colour and chlorine a very pale green colour. The colour in the bottom layer thus shows the less reactive of the two halogens involved in the reaction.

K. Work out

- Carry out displacement reactions with the halogens to find out more about the reactivity of the halogens. Draw grids of all the nine possible combinations (with chlorine, bromine and iodine) and predict what you think will happen before you start. Put a mark against those combinations that you predict will have a reaction.

  Plan the investigation carefully so that you can collect all the data you need.

  Keep a note of the results as you do the experiments.

- Look at the results from the displacement reactions and from the previous two experiments.

- Find out the boiling points of the halogens.

- Make a vertical list of the halogens, as they appear in the periodic table. Label the most reactive and the least reactive and briefly explain the reasons for your decisions.

  What two elements from group I and group VII would react most violently?

  What evidence do you have for this prediction?

L. Interpret

1 The seven most abundant elements in sea water are chlorine, sodium, magnesium, sulphur, calcium, potassium and bromine. Which of these are halogens?

2 Sea water is often called 'salt water'. Is this a good description? What evidence do you have for this?

3 Although there is only a little bromine in sea water (0.065 g in 1 kg of sea water) it is economic to extract it commercially. To do this sea water is acidified (to pH 3.5) and then chlorine is added. What will happen to the bromide ions that are in solution as the chlorine is added?

4 Assuming no wastage how much sea water is needed to get 1 g of bromine?

5 Actually 20 kg of water is needed to get 1 g of bromine. What is the percentage yield of the process in terms of bromine?

Section 5.4

MORE PERIODIC POWER

The periodic table is one of the most important aids for a scientist concerned with materials. You have looked at trends down the groups to the left and right of the table and in this section you will investigate the trends across the table. Use your knowledge to help you to make some predictions to begin with.

Try to find out about the elements and their compounds in the third period, that is sodium across to chlorine. Collect together all the information you can. Some of this you know

already, some you can find out by doing a quick experiment and for some you will have to use second-hand data. Your teacher will indicate which information you can find out by experiment.

M. Investigate

- Find out information about the elements of the third period and their oxides and chlorides. Use this information to complete the tables on Worksheet PB14 or Worksheet PB15 *Investigating period 3*.

The elements of period 3

11 Na Sodium 23.0	12 Mg Magnesium 24.3	13 Al Aluminium 27.0	14 Si Silicon 28.1	15 P Phosphorus 31.0	16 S Sulphur 32.1	17 Cl Chlorine 35.5	18 Ar Argon 39.9

You can use the results of this activity to find out if there are any *trends* across the period.

N. Interpret

- Use the results from activity M to answer the questions below. It may help if you begin each answer 'On the left......
  The first one is done for you.

  What do you notice about:

  1 the state of the elements; (*On the left they are solids, on the right they are gases*)

  2 how good the elements are at conducting electricity;

  3 where metals and non-metals are found in the period;

  4 the state of the oxides;

  5 the pH of the solutions of the oxides;

  6 the state of the chlorides;

  7 the pH of the solutions of the chlorides?

- Copy out period 3 as shown in the diagram and add arrows and notes to this diagram to help you to remember the trends. Make it as colourful as you can.

Extension

The trends that you noticed in the elements, oxides and chlorides in activity M are due to the electronic structure of the elements.

This equation shows how sodium forms an ion:
$$Na - e^- \rightarrow Na^+$$

Sodium readily gives up an electron to form the ion Na^+. It is a very good *reducing* agent.

Now go across to the other end of the period. You will remember that chlorine readily accepts an electron. This equation shows how chlorine forms an ion:
$$Cl + e^- \rightarrow Cl^-$$

As chlorine readily accepts an electron it is said to be a good *oxidising* agent.

These two elements, sodium and chlorine, readily form ionic compounds.

Silicon is in group IV. It has four electrons in the outer shell. When a compound is formed the silicon atom is unlikely to gain four more electrons or to give four electrons away. Hence it tends to form compounds with covalent bonding in which electrons are shared. It is, therefore, not very good as either an oxidising or a reducing agent, because it does not readily lose or gain electrons.

Two compounds of silicon: quartz (silicon oxide) and tourmaline (a complex silicate of aluminium and boron)

O. Interpret   **EXTENSION**

- Write down the electronic structure for each of the elements in period 3.

- Write a paragraph that explains:

  a how the electronic structure of the elements varies across period 3;

  b how this is related to the reactivity of the elements;

  c how a knowledge of this can be used when choosing oxidising and reducing agents for an experiment.

METAL EXTRACTION

Section 6.1

AN ENAMEL ENTERPRISE

Jewellery at a craft fair

Keith and Sue Cooke have been making enamel jewellery at home and selling it at craft fairs at weekends for some time. This has proved very successful and they are investigating the possibility of expanding this into a full-time business. Their major concern is how to cope with the increased amounts of acid that will be produced as waste.

They use sulphuric acid (1 part acid to 10 parts water) to pickle the copper. This removes the oxides from the copper prior to enamelling. They dip the copper shapes into the acid and leave them for a short time, until the copper goes pink, and they then remove them from the pickling bath. At the end of each week the acid needs to be removed. They plan to use the champlevé process of enamelling which also uses nitric acid (1 part nitric acid to 3 parts water) to etch a design in the surface of the copper. They hope that any system which is developed to cope with the waste sulphuric acid will also cope with the waste nitric acid.

Initially they expect to produce 10 litres of waste from the pickling process and 1 litre from the champlevé process weekly, but this may increase if the enterprise is really a success. They have drawn a simple sketch of a system that would suit them.

Their rough sketch

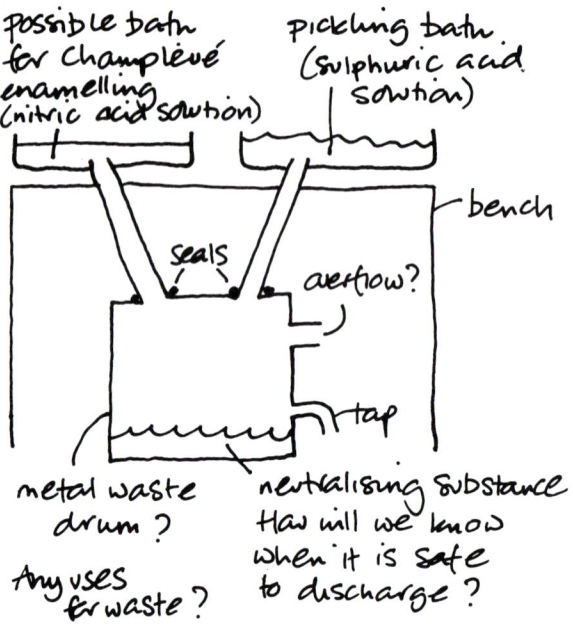

Before they commission the production of this system they need to know if it will work.

A. Work out

- The Cookes realise that they may not have thought of the best system and are open to alternative suggestions if their system can be proved to be not ideal. Your team has been called in to advise on the suitability of their system. Concentrate on investigating the following:

1   the best metal for the waste drum (copper, zinc, aluminium and tin have been suggested as possibilities);

2   whether the waste acids should be kept separately or mixed, as in the sketch;

3   what would be a suitable neutralising material for use in the drum.

- Produce a report for the Cookes so that they can decide whether or not to commission the system. Include in your report the likely cost of the materials for construction and also the best way of operating with it (for example, how frequently will it be necessary to empty the drum; what safety precautions will be needed and so on). If possible, give an estimate of the expected life of this system.

COPPER EXTRACTION IN TIMES PAST

Copper jewellery is not new. In fact it was worn about 6000 years ago in Asia. Copper was one of the first metals to be discovered, and pieces of native (naturally occurring) copper metal were collected and hammered together to form decorative pins.

Copper vessels made in Egypt about 2300 BC

Only small quantities of copper are found as pure metal. Most is found combined in minerals such as chalcopyrite or malachite. Minerals are found accumulated in rocks called ores.

Copper metal can be obtained from ores which contain the element by heating the ores in fires. Archaeological evidence indicates that this occurred near Lake Van in Eastern Turkey as long as 6000 – 7000 years ago. You can repeat this process now. All you need is some copper containing ore and some wood charcoal.

A sample of chalcopyrite found in Norway

B. Plan and test

- Try heating a sample of copper ore with wood charcoal to get copper.

  How will you know when you have obtained some copper?

- Draw a diagram of the equipment you used and indicate how it compares to what might have been available to the villagers of Eastern Turkey 6000 years ago.

- Write an equation for your process.

  In Unit 3 you considered different types of chemical reaction. What type of reaction is this?

Copper smelting based on this ancient process developed in many countries. Evidence has been found to indicate that planned smelting occurred in the Near East over 4500 years ago. Bronze weapons dating back to 1800 BC have been found in Britain, which indicates that another metal, tin, was also available to

A Cornish copper mine, around 1830

produce the alloy. Bronze is an alloy (mixture) of copper and tin. Copper ores are found in Wales, Cornwall and Cumbria.

A company called Mines Royal was founded in 1564 in England. One of the main aims of this company was for the Crown to benefit from any profit from mining. The three signatories to the founding of this company were Queen Elizabeth I, an English clergyman with an interest in technology and a mining agent. The company was to 'search, dig, try, roast and melt all manner of mines and ores of gold, silver, copper and quicksilver'. The queen was to have one tenth of the silver and gold and a royalty on other metals and 'the preferment in bying of all precious stones or pearl to be found in the working of these mines'. This resulted in a great row between the Queen and local landowners. The Queen was interested in the precious metals and the landowners were far more interested in the copper which by law would be theirs. A great trial was held between the Earl of Northumberland and the Queen over this. Records indicate that the judges decided that the Earl's mines contained more gold and silver than copper. There was no evidence for this, but such a miscarriage of justice was not uncommon at the time. It is not surprising that the Earl later supported Mary, Queen of Scots, in her claim for the throne.

Some of the shares in the company working the mines in Cumbria were bought by a German firm. About 50 'Almaynes' prospected for copper and set up their works in the Keswick and Coniston areas in 1565. They brought much needed employment for many local people, both in mining, timber cutting and charcoal burning. The area became prosperous and many schools were built which offered free education. Living standards rose.

The following extract from a poem describes how the charcoal burning was done:

The burners toil steadfastly,
setting sapling on the mount until it stands,
a passive lump, the wood entombed within the turf,
and wood among the trees around keening softly
as if it knows that each axe blow meant death for
each shorn stump.

At noon each day the waiting mound is fired.
Hot coals stream within its heart
and when the inner crack of singing wood is heard,
the final sealing turf is laid.
A muttering and crackling, the mound takes on life
that's all it's own, reeking smoke and steam
and uttering half heard thoughts
of boweres where the light runs green
and endless glades of sun
and forests by the sea.

All night and all next day the charcoal men
nurse the smouldering mound,
pouring earth on spurts of flame,
splashing water from the beck,
damping flames and catching sleep
as best they may, on bracken beds and turf.

For three long days and nights the smell of burning,
and then the fire seems to die.
The smoke is gone, but all there know
the embers still glow bright within.
And suddenly tired men regain their drive
as each mind the thought prevails:
Has it worked; has the burn gone well?
Slowly, the cooled mound is turned back.
The tired burners smile. The prize, the charcoal,
glistens satin black, and rings clear as a bell.

Irvine Hunt

C. Interpret

- Read the poem about making charcoal and then answer the following questions:

  1 What is charcoal?

  2 How was charcoal made?

  3 How long did it take the burners to make charcoal?

- Explain what the importance of charcoal was to the copper mines.

- Write an equation to show how copper ore is turned into copper during smelting.

- Name some items that were probably made from copper from these mines.

In Bougainville, Papua New Guinea, there was a development between overseas investors and the Papua New Guinea government. Just as mining brought improved living standards to Cumbria, so it did to Bougainville. The industry brought much needed revenue and employment to the area. Many years of planning occurred before any construction began on the islands. 80% of all the employees were local people and resources were made available to provide health, educational and leisure facilities. Unfortunately the local people became increasingly concerned about the exploitation of their resources and labour and, in 1992, this arrangement broke down.

Copper mining brought much needed revenue to Papua New Guinea and helped to improve the living standards of the people living there

Copper has been called the first useful metal. The demand for copper is still increasing. This is due to its many useful properties. It was the fact that it was quite a soft metal that could be easily worked that was one of its earliest attractions. This property and its attractive appearance make it popular with jewellery makers, sculptors and others. Copper is also the best conductor of electricity after silver. Electrical wires at home are made from copper. Another useful property is the fact that it is resistant to many forms of corrosion.

Copper has been alloyed with tin to form bronze and with zinc to form brass for thousands of years. You probably do not realise how much you rely on these alloys every day. They are used to make a wide range of items, from hooks and eyes to the casts that are used to make much equipment in industry. Other alloys are used to make numerous other everyday items, from coins to water pipes.

Copper(II) sulphate is another copper containing substance that you are familiar with.

This beautiful copper dome is part of the mosque in Regent's Park

D. Find out

- Copper has been used for over 6000 years. Find out some more uses of copper from the past and the present and think of some possible future uses of copper.

- Present these uses in a clear way and for each use state the property of copper that makes it suitable for this purpose.

MODERN COPPER EXTRACTION

About nine million tonnes of copper are produced each year. Only the production of iron and aluminium exceeds this.

E. Discuss

- In a group develop a list all the factors that a company will need to consider before developing an operation to produce a metal.

One of the main factors to take into account when considering metal production is how easily the metal can be got from its ore. This depends on how reactive the metal is. The most reactive metals are not easily extracted from their compounds. By looking at the reactivity series we can see how this affects the different methods of extraction.

The dates of discovery and the methods of extraction of various metals

Metal	Date of discovery	Method of extraction now
aluminium	1825	electrolysis of aluminium oxide
calcium	1808	electrolysis of calcium chloride
magnesium	1808	electrolysis of magnesium chloride
sodium	1807	electrolysis of sodium chloride
zinc	about 500 BC	electrolysis of zinc sulphate or reducing zinc oxide using carbon
iron	known since pre-historic times	reducing iron oxide using carbon
lead	known since pre-historic times	reducing lead oxide using carbon
copper	known since pre-historic times	thermal decomposition of copper sulphide using oxygen
tin	known since pre-historic times	reducing tin oxide using carbon

Typically copper is extracted from chalcopyrite ore containing about 1% copper but nowadays it is economic to extract copper from ores containing less than 1 part copper to 200 of ore. In the past these ores would have been left, as it would not have been viable to extract the copper. A number of factors are important in deciding if extraction is viable such as: the improved techniques for extracting the copper, the value of the purified copper and the value of the other metals in the ore. Frequently gold and silver are found along with the copper.

The extraction of copper has four main stages. What follows describes the extraction process at the Bougainville plant in Papua New Guinea.

A large, open-pit copper mine in Bougainville

1 Concentrating the ore

First the low grade ore containing on average about 0.5% copper is converted into a concentrate containing about 30% copper. The large mine in Bourgainville is an open-pit mine. The huge lumps of ore removed after blasting and drilling are loaded on to large trucks. Each of these trucks can carry 200 tonnes per load. The trucks empty the lumps of ore into a large crusher. Air and chemical reagents are added. Eventually the air forces the small particles of copper containing minerals to the surface in a process called froth flotation. The concentrate slurry is taken over the mountains to the shore. Here water is removed by vacuum filtration and the layer of copper containing concentrate is left. This is then exported.

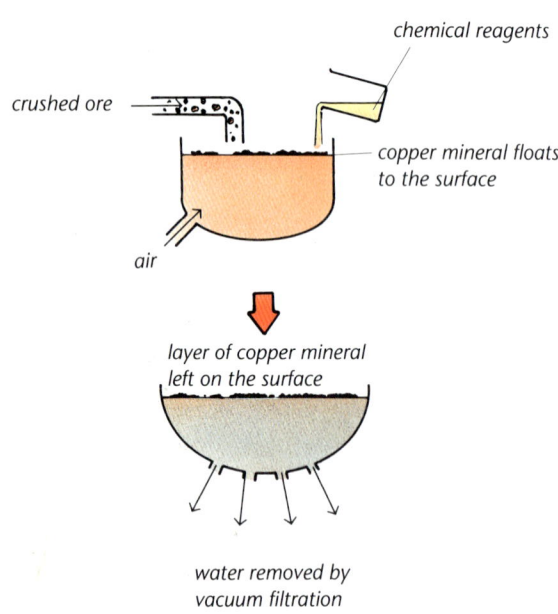

chemical reagents

crushed ore

copper mineral floats to the surface

air

layer of copper mineral left on the surface

water removed by vacuum filtration

2 Matte smelting

Here the ore is further concentrated. Sand is added and the sand and copper concentrate mixture is heated strongly at first. The process is exothermic. The heavy copper sinks and forms a matte containing 45% copper along with iron and sulphur. The lighter slag floats and is removed from the surface. The hot waste gas is fully used to produce steam (for power). Any copper dust is recycled and waste sulphur dioxide is used to make sulphuric acid.

3 Conversion

Air is blown through the matte as more sand is added. The iron is oxidised and with the sand forms slag which can be removed. The sulphur from the matte reacts with oxygen from the air to form sulphur dioxide. This is removed and is used to make sulphuric acid. The remaining copper, called blister copper, is 98.5% pure.

4 Refining

In this last phase the purity of the copper is increased to 99.99%. Initially air and then a hydrocarbon (a compound of hydrogen and carbon) gas is added to the copper as it is heated in a large furnace. This removes the blisters of copper sulphide. The copper is then formed into blocks, each weighing 300 kilos. Each block is used as an anode which is suspended close to a pure copper cathode in a large tank. The tank contains a solution of sulphuric acid and copper(II) sulphate. During electrolysis the remaining impurities in the anode (such as gold and silver) which are insoluble in the electrolyte fall to the bottom of the tank. Pure copper builds up at the cathode. The cathode, made up of copper which is 99.99% pure, is removed and rolled or extracted to make wire, bars, tubes and so on.

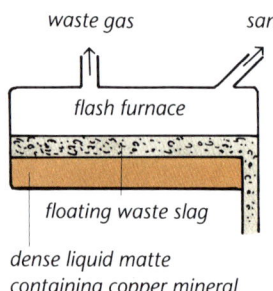

waste gas sand

flash furnace

floating waste slag

dense liquid matte containing copper mineral

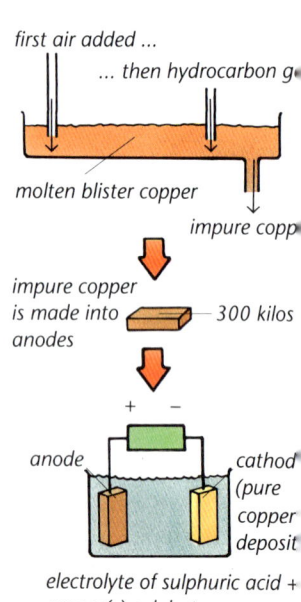

sand   matte at 1400k

sulphur dioxide

air

slag

blister copper

first air added ...

... then hydrocarbon gas

molten blister copper

impure copper

impure copper is made into anodes   300 kilos

+   −

anode   cathode (pure copper deposit

electrolyte of sulphuric acid + copper(II) sulphate

Micro-organisms are used to extract between 10 and 15% of the world's copper. Bacteria can feed on very low grade ores and release the copper in a soluble form. This process is likely to become more important in the future.

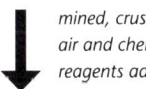

copper in chalcopyrite ore

mined, crushed air and chemical reagents added

copper in crushed ore floating in froth flotation vessel

copper in matte

F. Work out

- Use the information about copper production to answer the following questions:

  1 What raw materials are involved?

  2 What factors are important when deciding on the type of lorry to transport the copper ore?

  3 Why is the ore crushed before froth flotation?

  4 Why is vacuum filtration used rather than ordinary filtration?

  5 Why does a flash furnace not need any extra heat?

  6 Why is oxygen needed in matte smelting?

  7 Why does the slag float?

  8 What is the sulphur dioxide used for?

  9 How does electrolytic refining work?

  10 What chemical processes are involved in the production of copper by this method?

- Make a summary of copper production in the form of a flow diagram. Include as many chemical processes (and chemical equations) as you can. You could start with something like the one shown on the left.

G. Research

- Do some further research to answer the following questions. Try and write a paragraph about each one.

  1 Why does the producer bother to extract copper from ore that contains less than 1 part copper per 200 of ore?

  2 List the aspects of this process that could contribute to pollution. Suggest how each is, or could be, minimised.

  3 In 1977 it was predicted that the world reserves of copper would last for another 27 years. How can copper be conserved to last longer?

ALUMINIUM EXTRACTION

Although aluminium is the most abundant metal in the Earth's crust, unlike copper it is never found in its pure form. Aluminium has many important properties that contribute to its usefulness. It is very light and can be alloyed with other metals to produce very strong alloys. These are important factors, particularly when considering materials for the transport industry. Aluminium is a good conductor of both electricity and heat, and due to the development of a thin protective layer of aluminium oxide over its surface it is resistant to further corrosion. These factors contribute to its use in the building industry, the food packaging industry and the electrical supply industry.

Aluminium has proved difficult to extract from the clay in which it is found. In the 1880s it was regarded as a rare and expensive metal. This hardly seems possible now.

The extraction of aluminium has three main stages.

1 Mining the bauxite

The aluminium ore, which is called bauxite and is mostly aluminium oxide (Al_2O_3), is usually found near the surface. The top soil is removed and then the bauxite is loaded into trucks to be taken to the crusher.

Bauxite mining in the Provence region of France

2 Production of alumina

The bauxite is crushed and small lumps which contain little bauxite are removed. The remaining bauxite is ground to a fine powder before being reacted with sodium hydroxide under pressure. Aluminium hydroxide is formed. This sinks to the bottom of the tank and is removed by filtration. The aluminium hydroxide is then heated strongly to form aluminium oxide. This is called alumina and is a fine white powder.

3 Production of pure aluminium

The alumina is dissolved in molten cryolite (sodium aluminium fluoride) which is heated in a large furnace. During electrolysis the aluminium oxide is reduced to aluminium. This forms at the bottom of the electrolysis cell as the molten metal. The oxygen formed combines with the carbon anode to form carbon dioxide and carbon monoxide. The aluminium is made into large blocks, called ingots, which contain up to 4000 kg of pure metal.

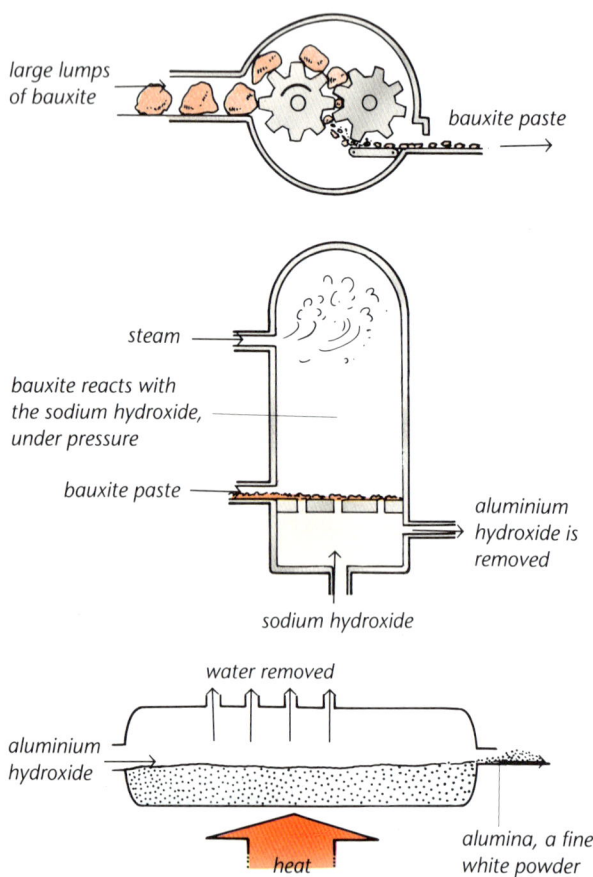

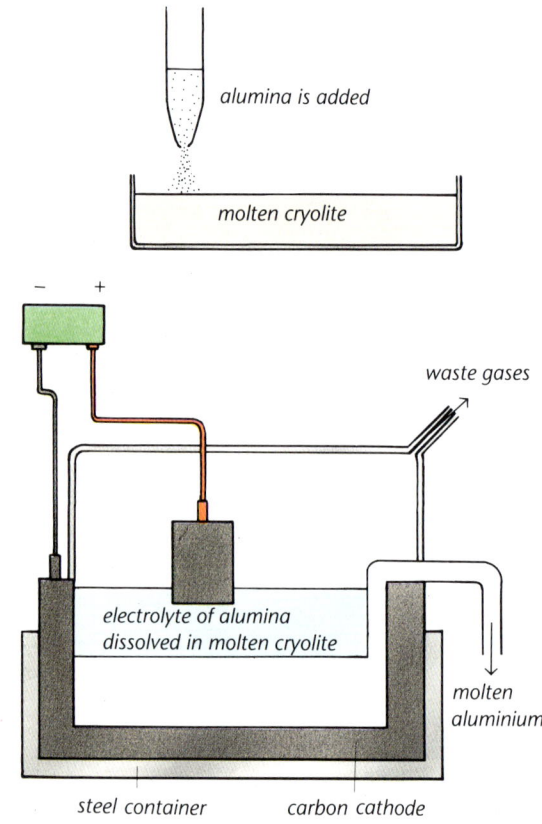

Stacks of aluminium billets at a foundry in Dubai

H. Work out

- Make a summary of aluminium production in the form of a flow diagram. Include on it as many chemical processes (and chemical equations) as you can.

- Draw up a table which shows the similarities and differences between the production of copper and the production of aluminium.

UNIT 7
USING MATERIALS

MORE MATERIALS

If you look around you, you can identify a large number of different materials. All materials consist of atoms joined together. The great variety of materials that exist is due to the fact that atoms come in different sizes and form bonds of different strengths and in different directions. The gross properties of each kind of material depend on this 'atomic building'. The uses of each kind of material is determined by its properties and these depend on the bonding and structure within the material.

It is now more and more common for materials scientists to be given a specification for a material's use. They then find a suitable existing material, adapt one, or frequently invent a new one for the job.

One important group of materials is the **metals**. Metals are such useful materials because of their characteristics. They are strong and yet can be rolled or hammered into shape. Above a certain temperature a metal becomes liquid. This enables it to be shaped by casting or to be combined with other metals to form an alloy. Metals can be welded. These characteristics and the ability of metal to conduct heat and electricity are due to their metallic bonding.

Glass is another important material. It was a popular material over 4000 years ago. However, the early examples of glass were opaque and usually coloured. Transparent glass objects were made by the Romans but what made glass transparent was only explained about twenty years ago. Glass does not easily fit into any of the three states of matter. It is an unusual solid in that its particles are not arranged in a regular way but are arranged as they are in a liquid. Glass is called a supercooled liquid. As liquid glass is cooled it turns hard but without forming the regular pattern of particles that would form in a crystal.

Glass is made from sand, sodium carbonate and lime. Glass made from just these three substances is used to make windows.

By incorporating other substances, such as boron oxide, sodium oxide and aluminium oxide, borosilicate glass is produced. This can withstand high temperatures and is used to make beakers for the laboratory and casserole dishes for cooking.

Two widely different uses of metals

Glass is now often used in the construction of buildings, as here in Milam in Texas

'Pyrex' glass can withstand high temperatures

Many materials are composites. A composite is a mixture of two or more different materials. The properties of some of these materials are shown in the illustration.

Glass reinforced plastic
This plastic polymer contains strands of glass fibre. It is light, like the plastic polymer but strengthened by the glass fibre. It is not as brittle as glass on its own.

Reinforced concrete
Concrete is the strong solid which forms when water reacts with a mixture of sand, cement and stones. This is good at withstanding compression forces where the concrete is being squashed together but when forces act to bend or stretch concrete it cracks easily. Steel rods are used to reinforce concrete so that it is less likely to crack.

Ceramics
Ceramics are mostly made from clay. They are hard and have high melting points. These properties are due to their structure. Strong bonds hold oxygen atoms in a tetrahedral arrangement around each silicon atom. As an oxygen atom can be part of more than one tetrahedra they can link together to form giant structures. The silicate tetrahedra contain a high proportion of oxygen – this makes them particularly good at resisting attack by oxygen and other chemicals. Ceramics are brittle, this is an unfortunate consequence of the strong covalent bonds within their structure.

Bone
The mineral hydroxyapatite part of bone is reinforced by the protein collagen. Hydroxyapatite gives the bone its hardness and the collagen gives it flexibility. The collagen has a similar function to the steel rods in reinforced concrete.

Fibres
Fibres whether natural or synthetic are polymers. They tend to be strong when pulled. This is due to their structure. The molecules are arranged in long ordered chains with a large amount of bonding to hold the chains in place.

Plastic polymers are the materials that are used to make many of the objects around you which, in the past, would have been made of natural materials. It is becoming easier and easier for scientists to design a suitable plastic polymer for a wide range of uses. For example, the waste pipe from the sink was traditionally made of metal. Now it is made from plastic, which is a lighter material. This may not seem important when you think of the function of the waste pipe, but some of the advantages are that it is cheaper to transport and it is also cheaper and easier to install.

A. Work out

- Work out how the properties of the composites shown in the illustration make them better for the jobs that they are being used for than the properties of the individual materials that they contain.

- Using the information in the text and in the illustration, and any other data available to you, make up a database or reference index that will help you easily to access information on the materials listed below:

reinforced concrete	glass reinforced plastic	ceramics
metal	laminate (plastic polymer)	nylon fibre
glass		

- For each of the materials include the following information:

  a name of material;

  b structure;

  c characteristic properties;

  d some uses.

  You may need to add to this later in the unit.